Fatal Flaws

FATAL FLAWS

HOW THE ASSUMPTIONS, MATERIALS AND METHODS INVALIDATE THE THEORY OF GLOBAL WARMING

Earl J. Seeley

Columbus, Ohio

Fatal Flaws: How the Assumptions, Materials and Methods Invalidate the Theory of Global Warming

Published by Gatekeeper Press
2167 Stringtown Rd, Suite 109
Columbus, OH 43123-2989
www.GatekeeperPress.com

ISBN (paperback): 9781662921469
eISBN: 9781662921476

This Book is Dedicated to the Memory of: Professor Clarence Ashton; Horticulturist, Brigham Young University Provo, Utah (USA)

Above Photo: From an adjacent mountainside I observed this structure seemingly framing a beautiful pine within the arch. After 5 days of exploring I finally reached this "arch". It required persistence, acceptance of risk, and Prof. Ashton would have said: "It was worth it." Along with hundreds of other students of Horticulture, Plant and Soil Sciences, I enjoyed countless hours of instruction from Professor Ashton: He lectured a little, stimulated interest with rhetorical questions, and asked frequently, "What do you think?". He would then listen, encourage further, deeper thought and expression, then with a knowing smile; encourage further thought, literature searching, and experimentation with the expression: "You are getting there, it's more complex than you think". And it always was.

The Significance of the Front and Back Book Covers

Front Cover: Water is the most prevalent substance on the face of the earth. Oceans and Seas cover 72% of the earth. When lakes, streams, and other surfaces which evaporate or transpire water are added, the effective percentage is even greater.

The prevalence of water and its amazing physical properties results in it buffering the earth's surface temperature, and providing an essential part of the conditions which favor the presence and viability of all living organisms. In addition, in conjunction with energy from the sun, it is the primary driver of weather in the short term and climate in the long term. As the atmospheric water content varies, weather and climate patterns change.

Back Cover: The sun adds color, beauty and heat to the earth's environment. The back cover illustrates how the added color and beauty enriches our lives. Without the sun's heat which is added to the earth's surface, the temperature of this orb would be approaching absolute zero (-460^{0}F).

The sun is the ultimate source of all the energy that allows life to exist. Its warmth and the photosynthesis it drives in plants provides our food, and most of the fuel which enriches our lives in many ways. Over the last 110 years the amount of sunlight energy reaching the surface of the earth has increased as the result of three processes.

First: The temperature of the sun at the quiet stage of each solar cycle has gradually increased since 1910.

Second: The number and intensity of sunspots and solar flares increased significantly during the 1940 to 1950 period and has remained historically high from then until 2018.

Third: The degradation of the atmospheric ozone over the last century has allowed a higher percentage of the suns energy to reach the earth's surface.

These facts about how the sun and water have contributed to the warming of the earth's surface over the last 110 years are completely ignored by the global warming theory.

Because of the assumptions and methods involved in the global warming doctrine these warming contributions are erroneously credited to the climate forcing of atmospheric carbon dioxide. There are other forces that have contributed to the warming of earth's surface temperature. The total contribution of these forces dwarf the contribution of the increase in the atmospheric carbon dioxide concentration.

I ask that you read this work with an open mind; then you be the judge.

About the Back Cover of this Book: My wife and I spent several years among the People of the Navajo, Hopi and Apache Nations. They have a very earth centered view of the world. The back cover photo is of "Spider Woman" in de Chelly Canyon east of Chinle Arizona. The perspective from which it is viewed, changes the visual impact of this magnificent structure, but the structure remains the same. Truth is truth. Seek it, refine it, test it, and continue the cycle until you know it.

About Earl J. Seeley, BA, MS, PhD (Age 78): He was born and grew up in an economically depressed area of Central Utah. He worked his way through college picking fruit, loading trucks at night, and pruning fruit trees. His goal in life was to become a basic scientific researcher. He attended Brigham Young University, majoring in Horticulture, minoring in Botany and Plant Physiology, with considerable chemistry, physics, soil chemistry, plant nutrition and climatology course work.

He attended Utah State University, completed his MS degree, then joined the USU Extension Service. This led to his employment as an Environmental Consultant with an Electric Company which was building generating facilities in central Utah. The company offered to pay his expenses to attend law-school, but his mind always wanted to argue both sides of an issue which does not fit the Socratic Method.

He returned to USU in the fall of 1974, completed his PhD and joined the Research and Graduate faculty at Washington State University in 1975. His research investigated the impact of air pollutants, cold, heat, frost, water stress, and light intensity on the photosynthetic capacity of fruit trees under laboratory and field conditions.

He was promoted to Associate Professor in 1980 but left WSU to join a private partnership in June of that year. He served as an Adjunct Professor at WSU from 1980 to 1986 and Brigham Young University during 1987 to 1999.

Since 1980 he has developed and managed several large agricultural enterprises, researched methods of frost control, plant/soil/water relations, plant nutrition, horticultural yield efficiencies, integrated pest management and has been a private agricultural management consultant.

Title, Purpose and Structure of This Book

The title of this book: "Fatal Flaws" refers to erroneous assumptions and methods which have been used to advance the theory of Global Warming. This book will define and examine the assumptions and science that underlies the global warming theory and demonstrate where and how they are flawed. This will be done as dispassionately as possible, without speculation as to motives or intent, as I believe those elements are beyond the scope of science.

Author's Note: In this book, I will sometimes use certain specific terms that are religious in nature to describe persons, concepts, conclusions and methods. I do this because, in my opinion, we are dealing with a secular religion which is approximately 38 years old, with foundational roots running back to the beginning of the environmental movement in the early 1960s. I would like to make it clear that my use of these terms is not meant, in any way, to denigrate persons or beliefs of any non-secular religion.

There are six main fatal flaws and three erroneous assumptions at the heart of the Global Warming theory, doctrine and dogma. Each is discussed in this book, and supporting evidence presented for the person(s) who desire to know the truth.

The fatal flaws occurred, when, in forming the "Theory of Global Warming," the contributions of four of the most important environmental factors in the control of earth's surface temperature were assumed to have had no impact on warming of earth's surface over the last 110 years. The remaining two fatal flaws concern carbon dioxide's roles and fate in nature.

The six fatal flaws are:

Fatal Flaw 1: The assumption that variations in the sun's radiant energy reaching the earth's surface, have had no impact on the increasing temperature of the earth's surface over the last 110 years.

Fatal Flaw 2: The assumption that the role of water, in controlling and modulating the earth's surface temperature and variations in the quantity of water in the atmosphere, has had no impact on the earth's warming over the last 110 years.

Fatal Flaw 3: The assumption that heat from the earth's core, in the forms of volcanic activity, magma flows, thermal vents, heat moving through the earth's crust and interactions between that heat and the earth's waters, has had no impact on the earth's warming over the last 110 years.

Fatal Flaw 4: The assumption that the contribution of increasing airborne particulate and aerosols to the greenhouse effect has had no impact on the earth's surface warming over the last 110 years.

Fatal flaw 5: The role of carbon dioxide's greenhouse impact has been vastly over-estimated. The increase in the level of atmospheric carbon dioxide since 1880 equals 121 ppm, or 1/83 of 1% of the earth's atmosphere. The total carbon dioxide content of earth's atmosphere equals 1/25 of 1% on a weight basis.

Fatal flaw 6: The dynamic usage and sequestration of carbon dioxide by the photosynthetic process, and three non-photosynthetic chemical processes has been ignored. Of the estimated 40 billion tons of carbon dioxide added to the atmosphere each year by human activity, at least 28.7 billion tons are removed (sequestered).

The goal here is to be thorough but brief, in a manner which is easily understood by the layperson who does not have extensive scientific training but wants to know the truth.

Authors Note: This work contains the following as appendix entries:

Page: 110 Appendix #1: The Normalization of Variance and Groupthink: This is my answer to the question: Why is the Global Warming theory so widely accepted, supported and believed?

Page: 116 Appendix #2: A Correlation Primer: A primer to help you understand what correlation means, what it is, and what is required to prove a causative relationship.

Page:118. Appendix #3: The Ultimate Fatal Flaw of the Global Warming Theory?

Read this Appendix Material as you feel the need.

Table of Contents:

Prologue

We are stewards of this earth and must solve the ecological problems our societies face and create or those societies will not survive. Before we can solve our environmental problems we must clearly define what they are, then properly allocate our resources to accomplish a true solution.

If we choose to solve the wrong problems we may spend a large portion of our present and future resources and find that we have only made the situation worse.

This work will present my science-based assessment, of the Global Warming Theory. The foundation of the theory will be closely examined including the main underlying assumptions. We will examine critically the accuracy of the science behind it. We will also examine the data manipulation needed and used to support the theory.

This book will challenge what many think they know about the foundational concepts of the Global Warming/Climate Change theory. I challenge you to read, contemplate and consider what is contained herein.

"It is much more complex than you think." There is a natural solution to the problem of rising carbon dioxide, to the extent that it is a problem. That solution is defined in this work.

Introduction

The global warming doctrine assumes that an increase in the earth's atmospheric carbon dioxide concentration from 290 parts per million (1880 level) to 411 ppm (7/5/2019 level) is the sole cause of any warming of the earth's surface temperature (foundational assumption #1).

The global warming/climate change theory has rested on this foundation for over 35 years. This is, scientifically, a ridiculous assumption as we will learn in the course of this work.

To some, the last sentence in the paragraph above may seem harsh and judgmental. The assumption identified above rests on a second invalid assumption.

That assumption is: The increase in the concentration of carbon dioxide in the air and the increase in the earth's surface temperature between 1880 and the present have been strongly correlated.

The second foundational assumption defined in the paragraph above is fatally flawed. The correlation between carbon dioxide concentration and the earth's surface temperature was strongly negative for the 30-year period of 1880 to 1910. This 30-year period constitutes what is referred to as a climatology.

We will discuss later in this work how this period of negative and other significant periods of weak correlation between the earth's surface temperature and the carbon dioxide concentration of the atmosphere constitutes a fatal flaw of the theory.

The third erroneous foundational assumption is that the correlation, strong or weak proves increasing carbon dioxide in the atmosphere is responsible for all the warming of earth's surface temperature that has occurred in the last 110 years. We will examine the above defined fatal flaws and assumptions in detail in this book.

With the claim in mind, that 100% of the increase in the anomaly (increased average global mean temperature) is due to increasing levels of carbon dioxide in the air, let me expose you to how the climate really works. It is a complex, intricate, and beautiful thing.

According to Lawrence Livermore Laboratories staff, in a peer reviewed article published in Nov. 2018, 40% to 50% of the ice melt in the Arctic over the last 38 years has been caused by natural variation in climatic patterns. This ice melt is traced to the El Nino cycle which is not human caused.

El Nino results in increased temperatures of the sea surface in the tropics, resulting in unusual high-pressure patterns around the poles. These high-pressure systems in the Arctic have caused a significant temperature increase and ice melt. Arctic ice melt is the linchpin of the global warming theory. The fact that it is not closely related to carbon dioxide forcing indicates a major fatal flaw of the theory.

We will discuss in this work the significance of this and other forces that have played key roles in the loss of significant amounts of the earth's ice volume.

Recent scientific findings have shown that about 6,000 years ago the Arctic rapidly lost its ice cover completely when the earth's surface temperature increased by 0.5^0 to 0.6^0C above "normal". This event, known as the "Holocene temperature conundrum" was related to events taking place in the tropics.

Throughout the history of the earth, environmental forces have been instrumental in driving increases and decreases in the earth's surface temperature. Global warming/climate change dogma insists that these forces have had no contributory impact on the slight warming of the earth's surface that has occurred during the last 110 years.

There are, in addition to the three fatally flawed foundational assumptions identified above, the six fatal flaws identified and briefly described on pages 6&7of this work.

Global warming advocates have, by ignoring the influence of these factors and ascribing all warming to a carbon dioxide greenhouse impact, grossly misled the disciples of the theory. Was it knowingly or was it the result of a myopic view of the environment that this happened?

It is not that difficult to comprehend. Albert Einstein is quoted as saying: "Make it as simple as possible, but not too simple". Attributing all warming of earth's surface to carbon dioxide forcing is an example of what he meant by the warning "but not too simple".

Chapter 1: The Little Blue Marble Deception

In Mr. Gore's first two books on global warming, he discussed in detail how a dedicated scientist had used 20,000 photos of the earth, taken from space, to show what the earth would look like if all the cloud cover was removed. He went on to opine that this view of the earth demonstrated how delicate the environment of the "Little Blue Marble" is.

This "Little Blue Marble" concept became a common expression repeated by politicians, climate scientists, news correspondents, NASA officials, on-line articles and global warming disciples. This idea became a major portion of the foundation of the global warming doctrine.

I was deeply impressed when I first saw this photo. When I read the conclusion reached by Mr. Gore, that this view of the earth demonstrated how delicate earth's environment is, I was instantly aware of the deception involved. (I will reserve judgment as to whether the deception was intentional or not.) I also realized that few people who read those words would recognize that deception for what it is.

There are two fundamental problems with the concept that the earth is delicate like a "Little Blue Marble". The concept results in a false, myopic, tunnel vision view of what this wonderful orb is: The problems are:

First: Viewing the earth from space, whether from 200 miles up or from the moon does not change the size, mass, or nature of the earth. The earth is not small. It is a vast, complex combination of interacting elements, substances and processes. The more one learns about it, the more intricate, complex, resilient, amazing and beautiful it appears.

Second: The earth does not exist in the "Little Blue Marble" state. That state is an artifact, created by meticulous effort to show the earth as it has never existed. Consider:

A. Approximately 72% of the earth's surface is ocean surface. When the surface areas covered by rivers, streams, lakes, glaciers, wetlands and ice deposits are added, the water covered surface is much higher than 72%. All these surfaces evaporate water, releasing water vapor into the atmosphere giving rise to clouds. On average more than 50% of the earth is covered by visible clouds.

B. Water vapor, liquid and ice in the atmosphere play a major role in limiting the amount of sunlight that reaches the earth's surface. Clouds reflect a significant amount of incoming sunlight back into space. They also absorb vast amounts of solar energy and then radiate it back into space. Without these functions (reflection, absorption and re-radiation of solar energy) the earth would be much hotter.

C. Water vapor is lighter than the other major components of the atmosphere (Nitrogen, Oxygen, Aragon, and Carbon Dioxide). This physical property of water results in water vapor moving from the surface upward in the atmosphere to an altitude where the temperature is cold enough for it to condense. At that point, the heat required to evaporate the water is released where it is easily radiated into space. In Chapter 7 we will examine the significance of this mechanism.

D. Water vapor, as it evaporates from earth's surfaces and rises into the mid-levels of the atmosphere where it condenses, is the driver which creates low pressure systems in the atmosphere. This function is key in modulating earth's weather in the short term and controlling climate trends in the long term.

E. The movement of water described in C and D above, carries 75% of the heat that is lost from the earth's surface into the mid-levels of the atmosphere. This heat is involved in the evaporation process, then moves with the water vapor as potential chemical energy to the altitude at which the water condenses. The energy involved by-passes a significant portion of the potential greenhouse effect of carbon dioxide and other minor greenhouse gases.

F. Because carbon dioxide is soluble in water, vast quantities are removed from the atmosphere as precipitation forms and falls to the earth.

G. Mr. Gore dismissed the role that atmospheric water, water vapor, and ice plays in regulating the earth's climate and temperature because it varies widely, changes rapidly, and is difficult to quantify.

We know a great deal about water; that knowledge deserves much more than a glib dismissal. True science is seldom easy.

Chapter 2: The Underlying Assumptions of the Global Warming/Climate Change Theory

An Allegory: A Myopic View of My Strict Diet

In 2009 I reached the point that I was a little on the chunky side at 5'10 and 215 pounds. Since I could no longer claim it was just my heavy bone structure or muscular build I began a strict diet regime. From March 1st to December 31st I began a weekly regime in which I added to my diet at least two and often three, 20 oz malts from a local mom-and-pop fast food outlet.

I also added an extra-large, high-quality ice cream on a waffle cone, as I made my weekly trip through the orchards I visit.

At the end of the year, after consuming over 100 large malts and 40 large ice cream cones, in addition to my normal diet, my weight had decreased to 186 pounds.

I hear you asking, what does this have to do with correlation analysis between the global mean anomaly and atmospheric carbon dioxide concentration? Just as the global warming advocates fail to acknowledge the role of the sun, earth's waters, heat from the earth's core and atmospheric particulate in the heating of earth's surface over the last 110 years, I chose to ignore one detail of my strict diet regime.

Beginning March 1st of that year I began to work toward my goal of hiking every side canyon along a 45 mile stretch of Utah Highway 31 between Fairview and Huntington Utah (USA).

There are 41 canyons that range in length from 2.5 to 11 miles with elevation gains from the start point to the summit of 1200 to 2800 feet. Three of those canyons require a three to four mile walk across rough country to get to the start point. Five of the canyons had some type of trail into, or along the bottom, the rest were as "rough as a cob" to use a rural expression.

Author's Note: The following interpretation is consistent with global warming advocates patterns and requirements. Their assumption is that 100% of the warming of the earth's surface temperature is attributable to the increase in the atmospheric carbon dioxide content.

If my assumption or hypothesis is that all my weight loss will be the result of my increased calorie intake, the following appears to be true.

If I did a correlation analysis between my calorie intake and weight during the 10-month period when I went hiking two or three times a week in rough terrain for four to eight hours, I find that my weight goes down even though my calorie intake goes up.

The above means that my correlation analysis will yield a negative correlation coefficient (r value), proving my weight loss is due to my increased calorie intake. This is the result forced by the underlying assumption that all of my weight loss will be the result of my increased calorie intake.

Just in case there are any doubts about how my strict diet regime is related to the basic point of this work, let me succinctly state it. Correlation is not proof of causation.

In my rather detailed description of my strict diet regime above, I related how my calorie intake was increased over the 10-month time period. I assumed that my weight loss was caused by the increased calorie intake (weight loss was negatively correlated to increased calorie consumption).

In actuality, the weight loss was caused by the increased calorie usage attributable to frequent, strenuous hikes which burned a lot of calories. In short, my initial inference (assumption) had no relationship to the true cause of my weight loss. My frequent, strenuous hikes did have a strong causative impact on my weight loss because the hikes were the cause.

The global warming/climate change, myopic, single dimensional model system relies on the assumption that all warming of the earth's surface is the result of climate forcing by increasing carbon dioxide in the atmosphere.

The assumption in the above allegory was that all weight loss was due to my increased calorie intake.

Neither of these assumptions are valid. Hence the hypothesis that my weight loss was the result of my increased calorie intake is without foundation and/or merit, as is the underlying assumption of the global warming theory.

I was taught, beginning with my parents, through my educational experience and by life, that when possible, avoid assuming anything. If you think you must assume something, examine the assumption as closely as possible to make sure it is reasonable. So, let us examine the facts and assumptions that are required to support the global warming/climate change theory.

The first foundational fact underlying the global warming theory is that the carbon dioxide concentration of the atmosphere has increased during the last 140 years. There was a gradual increase between 1880 and 1950 (total 22 parts per million). Between 1950 and 2020, the carbon dioxide content of the air has increased from 312 to 411 ppm. The total increase (1880 to 2019) has been 121 ppm which is equal to 29% of the current level.

Human activities have contributed to this increase, but other factors have also played significant roles as we will discuss later in this book.

The second foundational fact underlying the global warming theory is that since 1880 the earth's surface temperature has increased. The assumption is made that the increase is highly correlated with and caused by the rising carbon dioxide concentration of the atmosphere. This assumption is erroneous. Let us examine why.

First we must make sure we know how the earth's surface temperature is described. A system of expressions referred to as anomalies is used to describe how each year's global average surface temperature compares to a base (reference or normal) period.

The base period is the average annual global surface temperature for the period of 1951 to 1980. That 30-

year base period is used for at least two reasons.

First, the 30-year length of the period is the time span referred to as a "climatology" or the minimum length of time that is required to establish a climatic trend. Second, this was a period when the temperature was stable (it varied up and down, but the mean did not change greatly over time).

On page 21 is a copy of a graph from NASA which shows how the earth's temperature has changed over the years from 1880 to 2014. There are several noteworthy things about this graph.

The 0.0 line represents the mean global surface temperature for the years 1951 to 1980. This is the reference period mentioned above.

If the annual average global temperature is lower than the base period the anomaly will be preceded by a (-) sign. For example, the 1880 global mean temperature was, -0.16^{0}C. If the temperature was above the base line, for example in 1945, the figure would be 0.15^{0}C, the + sign being understood, not written. You now have a general understanding of the anomaly system. The annual mean temperature will be abnormally low, normal, or abnormally high.

Now let us discuss some of the important things this graph tells us.

First: At the left side of the graph (1880) you will notice that the lines start at, -0.16^{0}C and the general trend of the lines is downward to, -0.46^{0}C in 1910. That equals a 1/3 of 1-degree Celsius decrease. During this 30-year period, the carbon dioxide in the atmosphere increased by 11 parts per million (a negative correlation).

Second: Note that during the entire period covered by this graph there were frequent, significant, temperature swings up and down. The negative slope of the 1st climatology (30 years) and these significant swings of the global mean temperature up and down are extremely important. They demonstrate that something other than carbon dioxide is having strong impacts on the earth's surface temperature.

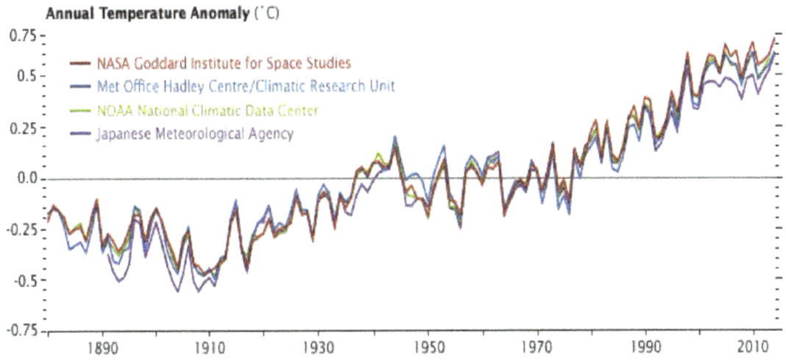

From: < https://earthobservatory.nasa.gov/world-of-change/DecadalTemp>

Third: Following 1912 until 1945 the global mean anomaly slowly rose to 0.20^{0}C, then for almost 40 years it varied widely, actually decreasing slightly. During the 100-year period (1880 to 1980) the carbon dioxide content of the air increased from 290 ppm to 338 ppm.

Fourth: The 30-year negative correlation plus the extremely weak correlation for the years between 1910 and 1980 means that the increasing carbon dioxide contributed very little to the warming of the earth. The 48.19 ppm increase in the carbon dioxide concentration (1880 – 1980) is equal to 40.2% of the total rise during the 140-year period from 1880 to July 2019.

Fifth: The temperature anomaly in 1880 was equal to, -0.16^{0}C but we do not know how far back the negative trend in temperature extended. We do know 2 things. 1. Before 1880 most of the earth's surface did not have any representative (thermometer based) temperature coverage and 2. That solar activity between 1610 and 1910 had been much lower than it has been since 1910.

As we will discuss in Chapter 6, this would support the conclusion that the negative temperature trend may have extended back to 1860 or further.

Next, look at the right side of the graph above and note how the lines move up and down between 1998 and 2011. The global mean anomaly varied over a narrow range during these 14 years while the carbon dioxide concentration increased by 24.88 ppm.

When the correlation between the change in the carbon dioxide concentration and the change in the global mean anomaly is analyzed, we find that the correlation was nearly non-existent during the time period 1998 and 2011 (it was actually negative before data manipulation: See Chapter 12).

During this period the content of carbon dioxide in the atmosphere increased by 25 ppm and the temperature remained the same. How does that happen if carbon dioxide is responsible for 100% of the increase in the earth's surface temperature?

The bottom line is: The assumption that the rising carbon dioxide concentration of the earth's atmosphere has caused 100% of the rise in the earth's surface temperature is a fatal flaw of the theory.

Some other force or forces acted together with the minor impacts of atmospheric carbon dioxide to control the earth's surface temperature during the 140-year period between 1880 and 2020. We will define what those forces were and how they are involved in the control of earth's surface temperature in Chapters 6, 7, 8 and 9.

Photo: This location in the Little Grand Canyon of the San Rafael swell in central Utah is a high elevation desert. The plant life is dependent on the precipitation that falls being directed into cracks and soil pits in the rock where they can exploit it during dry periods. It is an area that teaches one the true value of shade from any source, whether it be a rock, mountain ledge, or an occasional cloud.

Chapter 3: How Much Carbon Dioxide is in The Earth's Atmosphere?

How Has it Changed?

Most people do not comprehend the atmospheric carbon dioxide concentration numbers involved in the global warming theory. The two questions in the title and sub-title of this chapter, are a mystery to many.

The current level is 411 ppm, which is up from 290 ppm in 1880. To help us understand what those figures mean, let us envision a large bathtub which is 5 feet long, 2 feet wide and 2 feet deep. If we filled this bathtub to the brim with sugar we would need 96,000 teaspoons. Now let us imagine that this tubful of sugar is the earth's entire atmosphere, and ask the question: How much carbon dioxide is in there?

If we remove 3.84 teaspoons of sugar and replace it with 3.84 teaspoons of salt, the salt would represent (the carbon dioxide at) 400 ppm of the total contents of the tub. The ratio of sugar to salt, in terms of teaspoons would be 95,996.16 to 3.84. That is not much.

The next question would be: How much of the 3.84 teaspoons of salt represents the increase in the level of carbon dioxide in the atmosphere since 1880? The answer to that question is 1.152 teaspoons.

The bottom line is: The increase in the concentration of carbon dioxide in the atmosphere is not large enough to bring about the amount of heating that is claimed by proponents of the global warming theory.

Chapter 4: Some Things You Need to Know About the Global Warming Numbers

The first graph below is from Berkeley Earth(*p.107), one of the leading authorities on data sets, correction of data and the homogenization of data sets. It shows the plot of the annual mean global anomaly for the years of 1900 to 2014. I am placing this graph and the one shown previously close to each other to illustrate some of the fatal flaws of the global warming theory.

Examine the graphs on p. 26 closely, bearing in mind that these graphs plot the average mean global temperature anomaly for the same planet for nearly the same time period. The Berkeley Earth chart begins with 1900 and involves a different reference period (1900-1920), the other graph uses 1951 to 1980 as the reference or normal period.

One of the stated reasons for the development of the anomaly system was so there would be a widely accepted, reliable, standardized system for determining the earth's surface temperature.

Take a good look at these 2 graphs and consider the following 3 rhetorical questions: 1. Do the anomalies in the top graph closely parallel the anomalies in the lower graph? 2. Shouldn't a reliable system give a uniform pattern or result? 3. Are these graphs indicative of a reliable, standardized and accurate system?

Again, spend a little time examining and comparing these graphs. They illustrate fatal flaws of the scheme.

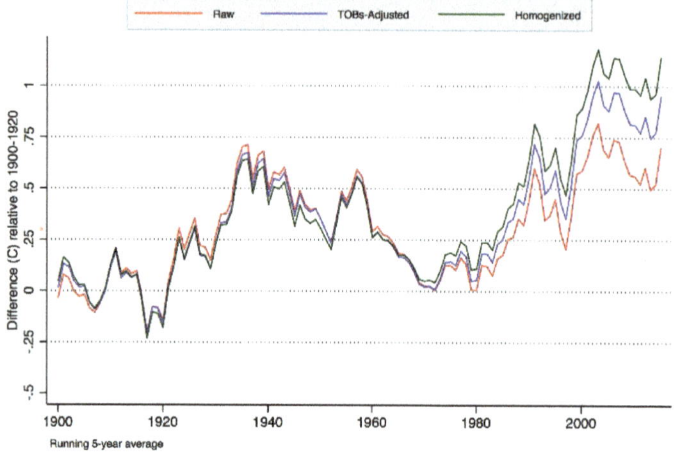

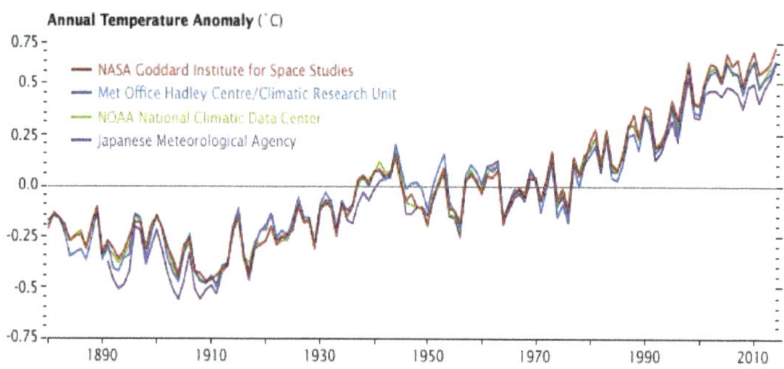

These plots are for the same planet and same general time period. So why is the 1935 global mean temperature anomaly between 0.65^{0}C to 0.70^{0}C in the top graph and between 0.0^{0}C to, -0.02^{0}C in the bottom graph?

These plots illustrate graphically one of the major fatal flaws of the scheme. When you want a larger (or different) anomaly figure adjust the start date or the reference period.

Rhetorical question: Should we conclude that the failure to develop a widely accepted, reliable, standardized system for determining the earth's surface temperature is a fatal flaw of the global warming doctrine? To help you with that answer; another rhetorical question: If I tell you that a very trusted source maintains that the 1935 average annual global mean anomaly was 0.70^{o}C would you trust that source?

Why is 1935 so different, but the above sources maintain that the 1979-1980 annual global mean anomaly was 0.00^{o}C? The atmospheric carbon dioxide concentration had increased from 309.7 ppm (1935) to 338.9 ppm, but the 1979–1980 global mean anomaly hovered near 0.0^{o}C on both graphs.

The Berkeley Earth plot shows a strong negative correlation between carbon dioxide and surface temperature for the 45-year time period 1935 to 1980. The other plot shows a strong negative correlation between carbon dioxide and the earth's surface temperature for the years 1880 to 1910.

One additional rhetorical question: Do you see why I maintain that there is little correlation between the atmospheric carbon dioxide concentration and temperature over time? Change the reference period, change the result.

The graph on p. 28, (The Berkeley Earth Graph) displays 3 different lines. The red line displays the raw (real) data. The blue line shows the corrected data and the green line represents the fully corrected and homogenized data. Take a close look at these 3 lines at the point where the red line descends to and touches the 0.5^{0}C line in 2011 (bottom of dip at far right).

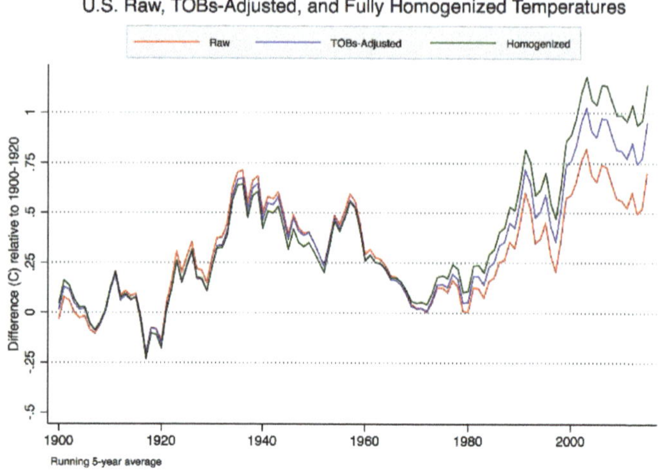

At this point the red (raw data) line shows the anomaly is equal to 0.5^{0}C after descending from approximately 0.8^{0}C in 2004. The blue line ("edited" data) indicates the anomaly was 0.75^{0}C a net increase of 0.25^{0}C over the real data. The green line (currently called the unadjusted data, the figures global warming advocates rely on) indicates that the anomaly at that point was 0.95^{0}C which is a net change of 0.45^{0}C above the real data after data manipulations were made. Remember this discussion, I will return to it in Chapter 12. Consider the following 4 rhetorical questions.

1. What justifies changing the raw data so that the anomaly is close to double that supported by the original data?

2. The result is a more doctrinally friendly number. Is that the purpose of the changes?

3. Is this science, or in the alternative, is it closer to witchcraft?

4. Beginning in the late 1970s there was a rapid conversion of temperature monitoring systems from thermometers to automated, durable, highly accurate electronic systems. Take a fresh look at the above graph and notice how, from the late 1970s to present the divergence between the three lines increases.

How can you justify the increasing divergence between the raw data and what the global warming disciples currently term "unadjusted data" during a period when monitoring systems became much more precise and accurate? You can't. This is a fatal flaw.

Now take another look at the three lines at the point the raw data line decreased to 0.5^{0}C in 2011. You will notice that from that point, all three lines began to rise rapidly.

What caused the shift from a descending trend to a rapidly rising global mean surface anomaly? During 2011 a new "master data set" (my term) was adopted. Not only by Berkeley Earth but by NASA and essentially all global warming advocacy groups. That data set is known officially as the GHCNv3 data set.

Prior to the adoption of this data set NASA had used a NOAA data set that had a significant, though not huge number and magnitude of data changes. Close examination of the changes made to that point in time showed most corrections were of "cooling biases". That means that each time a group of corrections were made, the net effect was to increase the alleged warming of the earth's surface.

The GHCNv3 data set added additional weather stations to the network and a 5-step data correction process. These corrections were made to the raw data to produce what global warming advocates call the "unadjusted data" or the data they rely on.

To me, using a 5-step data correction process on the raw (real) data and then calling the result the "unadjusted data" is, scientifically incomprehensible, and indefensible.

Before the GHCNv3 data set was adopted there had been 2 earlier versions, and the GHCNv3 data set was followed in May 2019 with a later version termed GHCNv4. The adoption of each of these different versions added additional cooling bias corrections (they increased the anomaly, and the projected long-term rate of increase in the purported value of the earth's surface temperature).

With thousands of observation points (29,000 in GHCNv4) and literally tens of millions of data points, the errors that resulted in cooling bias should have been very close in numbers and magnitude to the errors that resulted in warming bias.

That was not the case. The result defies basic statistical probabilities and it did so for each of the major groups of corrections. (a result that defies basic statistical probabilities can, theoretically occur, but not consistently, time, after time, after time, after time, after time, after time {six times in a row}).

The above discussion illustrates why changing data has been historically a forbidden practice in scientific endeavors. If the data acquisition system had been properly designed, calibrated, installed and maintained there would be no need to resort to suspect, scientifically indefensible data corrections.

These corrections can only be viewed as rampant bias, or the result of poor scientific design of the "experiment". Several of my mentors referred to this type of shoddy scientific procedures as "freshman mistakes". Neither of these possibilities can be viewed as complimentary of those involved.

The most troubling and revealing part of the above graph and discussion are the excuses used to justify the changes. As pointed out above, the number and magnitude of the required changes increased substantially between 1980 to 2019. The equipment used was increasingly accurate, durable, and reliable. The question remains: Why are the changes necessary? They are not justifiable!

As a youngster who grew up in cowboy country. I considered it a privilege to have known one of the last true western cowboys. Burl Majors claimed to have read one book in his life, a volume of the works of Shakespeare which he carried in his saddle bags (and after a little to drink he would quote from that source extensively).

He had lived in desert country so was not a fan of fishing. I never heard him say something smelled fishy, but I did hear him describe some of the most offensively odoriferous occurrences in nature as smelling worse than a 2-week old Limburger Cheese and Salami sandwich which had been stored in a warm saddle bag. I think that applies to this situation.

The true increase in the earth's surface temperature lies somewhere between $0.45^{0}C$ and $0.55^{0}C$, not the $1.1^{0}C$ that has recently been adopted.

Because of a lens attachment, this photo reveals only a part of the full picture. The tall structure on the left is one of the Castles that give Castle Valley Utah (USA) its name. You probably will never see the full picture because of the myopic (limited) view of this photo. Hopefully the next chapter gives you a complete view of the roles and fates of atmospheric carbon dioxide.

Chapter 5: Carbon Dioxide; It's Roles and Fate in Nature

Carbon dioxide is one of the most important substances on this planet. Without it, life could not exist. From viruses to mammals all are dependent on carbon from carbon dioxide for the structure, maintenance and energy of their life status. This gas plays several important roles in nature. As it performs these roles, a dynamic flow of utilization, production and sequestration results.

Global warming doctrine focuses on a bit role of this important actor, its minor role in the greenhouse effect. The doctrine entirely ignores the leading roles played by this gas. In fact, it is viewed as a dangerous pollutant by many global warming advocates.

One of the most important leading roles atmospheric carbon dioxide plays is as a substrate in photosynthesis. This process, by which green plants utilize some of the suns energy to combine carbon dioxide and water to form sugars, amino acids and other building blocks of life uses 706 billion tons of carbon dioxide from the atmosphere annually.

That is 17.65 times the amount of carbon dioxide released annually into the atmosphere as a result of human activity.

Much of the 706 billion tons of carbon dioxide is returned to the atmosphere by short term respiration processes of plants and animals. It is also returned to the atmosphere by decomposition of organic materials by oxidation, bacteria and fungi. A significant amount of the carbon dioxide used in photosynthesis is sequestered for short, medium and long terms.

In addition to photosynthesis there are three very important physical/chemical process that remove carbon dioxide from the atmosphere and sequester it for very long periods (much of it is sequestered for thousands, even millions of years). We will discuss all of this and more in this chapter.

The efficiency and rate of the photosynthetic process is increased by the higher concentration of carbon dioxide that now exists in the atmosphere. This increased efficiency has played a significant role in increasing food and fiber production to support the needs of an expanding population.

It also benefits forests, grasslands, wet lands and other plant communities that use photosynthesis for the basic building blocks of their growth and development.

As an example of how increased carbon dioxide concentrations can increase food production consider that greenhouse vegetable production facilities often raise carbon dioxide levels to 1,000 to 2,000 ppm. At that level, with nutrition, temperature and light conditions optimized, 1-pound lettuce heads can be produced in 25 days, plus the 7-day period necessary to produce a healthy transplant. At atmospheric concentrations typical of the late 1960s (320 ppm) it takes 42-48 days for the process.

A second example of how the higher carbon dioxide levels in the atmosphere are beneficial to the environment is that they support 17% to 25% higher photosynthetic rates in most hardwood and softwood species of forest trees. These trees can sequester vast quantities of carbon dioxide in well managed forests.

During five to six months of the year, carbon dioxide is removed from the atmosphere faster than it is added. That may surprise many, but for the doubters you need go no further than the book "An Inconvenient Truth" pages 26 and 27. Mr. Gore shows the carbon dioxide concentration of the atmosphere for the years 1958 to 2005 (The Keeling Curve).

The red line goes up and then goes down, once each year. The ascending portion of this line represents the portion of the year when it is winter in the northern hemisphere and the total, world-wide releases of carbon dioxide are greater than the total utilization. The descending portion of the line represents the summer period in the northern hemisphere when the utilization of carbon dioxide, world-wide, exceeds its release.

This pattern of annual changes in the atmospheric carbon dioxide concentration suggests a natural solution to whatever problems the increasing carbon dioxide concentration may present. We will discuss that solution in chapter 11.

Another highly significant beneficial effect of the faster growth rates of crop plants under the higher carbon dioxide concentrations in the atmosphere is that it saves water. Because the photosynthetic rate is increased and water usage remains very close to constant, it requires less water per pound of product produced.

In addition, there are three non-photosynthetic processes that occur 24/7/365 and sequester carbon dioxide from the atmosphere for long periods. Most of this sequestration will last for millions of years.

Carbon dioxide is constantly being removed from the air as it dissolves in liquid water in raindrops, seas, lakes, streams, and fog. When carbon dioxide is dissolved in water a portion of it goes through a molecular change as it reacts with water to form carbonic acid.

Some of the carbonic acid will then break down into carbonate or bicarbonate. These products increase in earth's waters as the concentration of carbon dioxide in the atmosphere increases.

Once formed the carbonate and bicarbonate may persist in solution or combine with basic elements such as calcium and magnesium to form complexes which eventually become limestone and other forms of sequestered carbon dioxide.

The basic elements involved in these reactions with carbonates are calcium, magnesium, potassium and sodium. There has been a good deal of speculation that the amount of material involved in these reactions may deplete the basic elements in the sea and result in a sudden steep increase in the atmospheric carbon dioxide levels. This speculation is related to what is termed, acidification of sea water.

The advocates of this theory ignore two very important facts. First, the vastness of the seas and other bodies of water results in huge amounts of carbon dioxide constantly dissolving in the earth's water. Some of this carbon dioxide then forms carbonic acid which reacts with rocks, rock particles and even soil to release additional basic elements. It is a process that takes place constantly. The higher the carbon dioxide level in the air, the faster this process takes place.

Second, the amount of dissolved carbon dioxide, carbonic acid, carbonate and bicarbonate in the seas is estimated to be 36,000 Billion tons.

The third group of chemical processes that result in sequestration of carbon dioxide from water, are the processes that result in formation of soil from rocks. These are very important reactions and have been ongoing for three+ billion years. They have played key roles in making the earth such a wonderful place to live, work, and play.

If you would indulge me I would like to introduce you to one of the miracles of nature. It is something that has interested me for about 70 years as I have hiked, ridden horses, and traveled the backcountry (also as I have accomplished research in the laboratory and field). I had not thought of it as a mechanism that sequesters carbon dioxide until about 20 years ago. It is very important and takes place in billions of places all over the world and lasts for ages.

Authors Note: And some of you thought all carbon dioxide does is give your soda a little fizzy bite.

Plants use photosynthesis to form carbohydrates and

organic acids. A significant amount of this material is moved from the leaves to the roots of the plant. In the roots the carbohydrate is used as an energy source to support root growth and to provide materials to build and maintain the root system.

Respiration produces energy to support the life processes in the root system. Water and carbon dioxide (by-products) are released into the root environment where they may react to form carbonic acid. Many plants exude other organic acids from the roots.

These acids break down rock, sand and soil particles in ways that provide a better root environment in terms of mineral nutrients, soil structure and even space for root development. The process usually begins with the action of lichens, the pink, gray and blue green structures that cover the rocks in these photos.

Next comes the moss which can be seen underneath the grasses and broadleaved plants in the "pot" in which these plants are growing. The yellow spots on the moss are the fruiting (reproductive) structures. At the top left there are some liverworts which are, like the lichens and moss, primitive forms of plant growth.

When the moss and liverworts have made enough soil, the grasses and broadleaved plants can become established. As the process continues it creates more space and soil so larger plants may become established and continue the process at an accelerated pace.

The photo on the left shows a larger plant community that has developed to the point that several small evergreens have become established. As the small pine trees grow, they and the associated plants, fungi and bacteria will break down more rock, forming more soil. The roots of the pine trees will grow into cracks and fissures in the rocks and literally split them. In the resulting space additional root growth and soil formation will occur breaking down more rock, creating more space.

This carbon dioxide-driven process takes place 24/7/365 in billions of places on this big wonderful orb. This becomes a perpetual cycle as plants utilize carbon dioxide in photosynthesis to make sugars, amino acids, cellulose and other elements of their lives.

The plant root systems, and the carbon dioxide and organic acids they release into the soils, break down additional rock and rock fragments. This releases mineral elements that nourish the plants. Some of the released mineral elements react with the carbonate and bicarbonate to form insoluble precipitates.

The result is the chemical sequestration of the carbon dioxide as it forms organic matter and precipitates within the soil and rocks. The precipitate deposits are like those found in coral, seashells, and the deposits that form on your bathroom fixtures and shower stalls.

We hear over and over what a threat carbon dioxide accumulation in the atmosphere is to our way of life and to our very existence. We seldom hear what carbon dioxide does to benefit us. We also never hear that there are processes that remove from the atmosphere and sequester most of the 40 billion tons of carbon dioxide that is added to the atmosphere on an annual basis by human activity.

It is an undisputed fact that carbon dioxide is being added to the atmosphere faster than it is being removed. The concentration in the air is increasing by approximately two parts per million per year. That amounts to approximately 11.35 billion tons per year not the 40 billion tons released by human activity. The 28.7-billion-ton difference is because of the sequestration brought about by photosynthetic and chemical sequestration processes.

In Chapter 10 we will discuss how changes in wildfire control methods and procedures have resulted in a significant increase in the rate of atmospheric carbon dioxide accumulation.

In Chapter 11 we will examine how deforestation of approximately 1 billion hectares (2.47 billion acres) of earth's land surface since 1920 has played a major role in the increase of this gas in the atmosphere.

But first let us discuss what is causing most of the earth's surface warming.

The assumption that an increase of 121 ppm in the carbon dioxide content of the atmosphere is solely responsible for all the warming of the earth that has occurred over the last 110 years is preposterous. There are other environmental forces which are much stronger, prevalent and active in regulating and modulating the earth's surface temperature.

These forces are in order of importance: 1. Sunlight, 2. Water, 3. Heat from the earth's core; and 4. Increased particulate (aerosols) in the air. In the aggregate the impacts of these four forces dwarf the impact of the 121 ppm increase in the carbon dioxide concentration in the atmosphere, part of which is attributable to human activities.

These forces vary over time and have contributed to the surface and near-earth atmospheric temperature increases that have occurred over the last 140 years (to be technically correct the temperature increase has been limited to 110 years due to the temperature decrease that occurred between 1880 and 1910).

In the third paragraph above I stated: The assumption that an increase of 121 ppm in the carbon dioxide content of the atmosphere is solely responsible for all the warming of the earth that has occurred over the last 140 years is preposterous.

The central reason that this statement is true is that the assumption involves a single dimension interaction, consistent with the single dimension models used in climate science. A single dimension model cannot describe the net interactions of the five main factors involved.

Let me describe as precisely as possible how a single dimensional interaction (model) works and contrast that with a multi-dimensional interaction.

Viewed through the single-dimensional model between temperature and carbon dioxide, with the basic assumption that, “Any heating will be the result of carbon dioxide forcing” let us examine the contents of a mason jar.

We begin with a 1-quart mason jar containing 1 pint (473 g) of water at a temperature of 20^0C.

A. We add a net of 473 calories of heat as radiant energy from carbon dioxide.

B. The result is that the water now has a temperature of 21^0C.

C. The conclusion: Carbon dioxide has caused a 1^0C rise in the water temperature.

D. We now add a net of 473 calories of heat as sunlight energy.

E. The result is that the water now has a temperature of 22^0C.

F. The conclusion: Carbon dioxide has caused a 2^0C rise in the water temperature.

Does the above make sense? Answer: No. But, remember the controlling assumption: all heating will be the result of carbon dioxide forcing of the contents of the mason jar, hence the conclusion, no matter how ridiculous; the heating is all the result of carbon dioxide forcing.

If the heat came from water, magma, or the greenhouse impact of aerosols the conclusion would be the same, forced by the underlying assumption. The heating is all the result of carbon dioxide.

Viewed through a multi-dimensional model between temperature and carbon dioxide, with the basic assumption that any heating of the contents of the mason jar may be the result of energy from the sun, water, heat from the earth's core, aerosols, and/or carbon dioxide, we see the following:

We begin with a 1-quart mason jar containing 1 pint (473 g) of water at a temperature of 20^0C.

A. We add a net of 473 calories of heat as radiant energy from carbon dioxide.

B. The result is that the water now has a temperature of 21^{0}C.

C. The conclusion: carbon dioxide has caused a 1^{0}C rise in the water temperature.

D. We now add a net of 473 calories of heat from sunlight energy.

E. The result is that the water temperature is now 22^{0}C.

F. The conclusion: carbon dioxide has caused a 1^{0}C rise in the water temperature, and the sun has caused a 1^{0}C rise in the water temperature.

If the heat added in A, or D resulted from setting the jar in warm water, or on cooling magma, or from the greenhouse effect caused by aerosols, the heat would be credited to the proper source. That does not happen with a single dimension, myopic modeling system. It is all lumped together under the carbon dioxide forcing moniker.

In the more comprehensive version of this work entitled "Fatal Fails: Scientific Problems with Global Warming" to be published 4/1/2022, in eBook and paperback formats, you will find examples of how single dimension models fail when applied in nature.

Chapter 6: The Sun's Role in the Control of Earth's Surface Temperature

This photo: (2/11/2021) shows the surface of the quiet sun. Since early 2018 the sun has had very few or no sunspots and solar flares. This has caused a significant cooling of earth's atmosphere and surface. This cooling has been monitored by SABER, a spectroradiometer on a satellite monitoring the temperature of carbon dioxide molecules in the upper atmosphere. This phase of the current solar cycle is highly significant because of what it tells us about the foundational assumptions of the global warming theory. (spaceweather.com Credit: SDO/HMI) In the more comprehensive version of this work entitled: Fatal Fails: Scientific Problems with Global Warming, you will find precisely how this cooling occurs.

The sunlight that reaches the earth is, without question, the most important factor in determining the earth's surface temperature. The distance of the earth from the sun, and the tilt of earth's axis, which determines the seasons, are nearly constant. Small changes in the temperature of the sun will cause variation in the sun's impact on the earth. So the question is, does the sun's radiant energy that reaches earth's surface vary?

The answer to this question is, yes it does. Forces within the sun result in cyclic increases and decreases in the number of sunspots and solar flares. Changing the number of sunspots and solar flares changes the temperature of the sun, which changes the amount of energy emitted by the sun that reaches the earth.

The period between the beginning of the increasing portion of the curve and the beginning of the next increase is referred to as a solar cycle, roughly 11-13 years.

Beginning in the 1880s there was a gradual increase in solar activity as you can see in the plot below. In the 1940s there was a major increase in solar activity.

This chart is part of an effort made to determine the amount of energy that reaches the outer limits of the

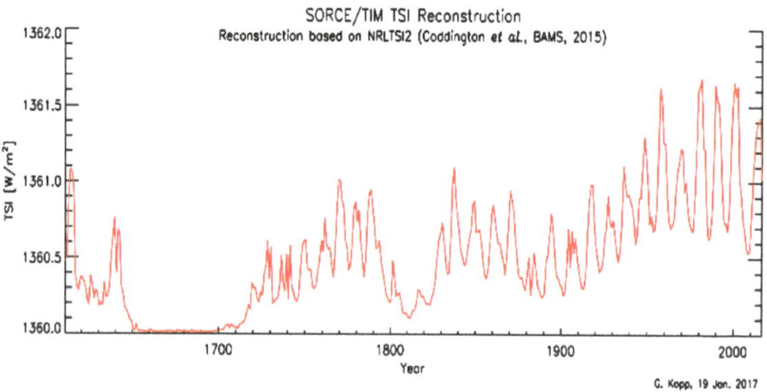

earth's atmosphere. This recent work is unique because it defines what has happened to the temperature of the sun during the "quiet sun" phase of the solar cycles, as well as during the active portion of each cycle.

The narrow peaks represent the rise and fall in the numbers of sunspots and flares. The area underneath the base of individual peaks shows the intensity of the sun light (irradiance) that reaches the earth's atmosphere during the quiet sun phase.

Notice how the area underneath the base of the peaks gradually increased between the 1880s and the early 2000s. This is the intensity of the sunlight that is received by the earth's outer atmosphere during the quiet stage of the cycle (in watts per square meter).

Since 1810 the intensity of the solar irradiance reaching the outer atmosphere has varied from 1360.1 to 1361.7 or by a total of 1.6 watts per square meter.

The solar cycle represented by the partial peak at the extreme right of the chart, is cycle #24. The higher base level during the last 10 solar cycles is enough to increase the energy that reaches earth's surface by about 15% of the excess energy retained by the earth during the last 110 years.

I apologize for getting a little technical, but what this chart illustrates is a key point that demonstrates how wrong the global warming theory is when it maintains that all the heating of the earth's surface is attributable to the greenhouse effect of carbon dioxide.

The above chart shows:

First: The increased solar irradiance from the quiet sun accounts for 15% of the excess heat that has been retained by the earth's surface during the last 110 years.

Second: Notice the highly significant increase in the height of the last 7 peaks. The increase in the average solar irradiance attributable to the intensity of the active portions of those solar cycles is adequate to account for an additional 45% of the excess heat absorbed by earth's surface during the 1880 to 2019 period.

These two impacts have caused a major portion of the warming of the earth in recent decades. But wait, there is a third way the sun has been instrumental in increasing the earth's surface temperature.

In the mid-1970s we became aware that the ozone in the upper layers of the atmosphere was being depleted. It was determined that this depletion was due to Chlorofluorocarbons (CFP's). As the ozone was depleted, the amount of X-ray and ultraviolet light reaching the earth's surface increased (a warming impact).

The offending materials were used primarily as refrigerants and propellants, and the use was worldwide. Two UN sponsored treaties were signed by 196 countries and the European Union in 1985 and 1987 to deal with this problem. The treaties phased out the use of these and other ozone depleting substances.

The ozone in the atmosphere has been slowly recovering because of the natural ozone formation processes since 2005. It will take years for the warming impacts of this factor to be totally negated.

Recent, authoritative, peer reviewed research has estimated that 50% of the warming of the arctic has been due to depleted ozone levels in the atmosphere. This is a human caused impact, but it is important to remember two things.

a. This warming has nothing to do with the greenhouse effect caused by carbon dioxide and other minor greenhouse gases.

b. This warming factor has been effectively dealt with and we will see a gradual re-establishment of the ozone levels in the upper atmosphere which will diminish this warming impact.

The above three factors that affect the amount of sunlight that reaches the earth's surface have accounted for much of the warming that has been experienced in the last 110 years. This warming of the earth's surface has had nothing to do with the atmospheric carbon dioxide content.

Should I repeat that? I won't, because my goal is to write this book with 25,000 words or less.

The global warming/climate change theory, because of its myopic, tunnel vision, single dimensional approach, maintains that these impacts of the sun do not exist. Instead, it attributes all the sun's contributions to the earth's surface warming, to the greenhouse impact of carbon dioxide and other minor greenhouse gasses.

This fact constitutes the most important fatal flaw of the global warming theory. A much more detailed treatment of this topic is contained in: Fatal Fails: Scientific Problems with Global Warming: To be published 4/1/2022.

Some readers have undoubtedly noticed that in the last few paragraphs I have noted three items that have accounted for 15%, 45% and 50% of the total amount of excess heat that has been retained by the earth's surface.

You have also remembered that there have been other things that I have mentioned, and there are more that I will recite, which contribute to the earth's warming. For example, the El Niño impact on the arctic temperatures has caused 40% to 50% of the Arctic ice loss that has occurred during the last 38 years. You are right, the total adds up to significantly more than 100% without the contribution from earth's core heat, atmospheric particulate, and changes in the amount of water in the atmosphere. How can that be?

I am glad you asked. Each of the above questionable percentage figures are the product of single dimensional modeling efforts. This is the basic approach that underlies the global warming doctrine. As I have suggested, and will say several more times, this system distorts the view of the environment because it does not consider the complexity of nature.

The increased sunlight energy warms the surface. The warmer surface evaporates more water. The water vapor rises and carries the heat required to evaporate it to elevations cool enough for it to condense. The warmer the surface, the faster this process moves heat into the mid-levels of the atmosphere. From this point, most of the heat is radiated into space.

This is an example of how the 4 major forces that regulate the earth's surface temperature interact to buffer and control the earth's weather and climate. In this manner, each of the warming forces are buffered and a significant amount of the excess heat is moved away from the earth's surface. Nature is amazing, beautiful and intricate. It is impossible to describe with a single dimensional model.

This photo was taken west of Polacca Arizona on the Hopi 3rd Mesa. It shows a phenomena which translates into English as "Sun Dog". When you see this, you can count on 2 to 3 weeks of unsettled weather. We have experienced at least a year of unsettling science, for each day this sun dog forecasts unsettled weather. As you will see in this work, we have experienced 2½ decades of unsettled and unsettling science since global warming became a popular subject .

Chapter 7: Water's Role in The Control and Modulation of Earth's Surface Temperature

Water is the second most important factor in the control of earth's temperature, exceeded only by the sun and its influence. It is a feedback (buffer against temperature change), a role acknowledged by global warming advocates. It is also a primary driver of weather and climate, a role ignored, even denied, by the disciples and doctrine.

In late evening, this setting makes great Bonneville Cutthroat trout fishing. I am an ex-fisherman who now enjoys hiking and observing nature. If one of my grand-children would like to learn to fish I can quickly revert to former ways.

When Mr. Gore glibly dismissed the role of water in his first books on global warming, he established "the word" for global warming/climate change disciples which has been accepted on blind faith. It could not be more erroneous. In dismissing water's contribution to the control of earth's surface temperature, he did two things.

First: He assured that the global warming movement would be established on a foundation of quicksand.

Second: He insured that any warming influence that water contributes would be credited to carbon dioxide forcing of earth's temperature. An example of how that happens was included earlier when we discussed how 40% to 50% of all the ice loss from the Arctic ice pack has been caused by natural climatic variation traceable to El Niño impacts.

El Niño is a sea surface (water) temperature driven phenomenon. It is part of what scientists from Lawrence Livermore Laboratories refer to as natural variation, i.e., it is not human caused.

How many times have you heard the apocalyptic pronouncement, "The Arctic ice is melting, and it is all because of rising atmospheric carbon dioxide?" Unless you live deep in the woods with no devices, a good estimate would be once a day for 30 years (equals 10,950 times).

Water is a unique substance. Its chemical and physical properties set it apart from all other substances on this earth. These chemical and physical properties, when combined with its vast quantity and distribution, makes it an important driver and modulator of earth's weather and climate.

Some of the extremely important roles water plays in the control and modulation of earth's surface temperature are:

1. Water evaporates from earth's surface and then rises into the mid-levels of the atmosphere where it condenses into clouds and precipitation. This process transports approximately 75% of all the heat that leaves the earth's surface upward in the atmosphere. From the point that it condenses, the released heat is more easily radiated into space. This process by-passes much of the potential greenhouse impact of carbon dioxide and other minor greenhouse gases. This impact is ignored by the global warming doctrine.

2. Water vapor is much lighter than atmospheric nitrogen, oxygen, argon and carbon dioxide. Because of this physical property, it is the driving force behind changes in atmospheric pressure. It is the physical force that causes high-pressure and low-pressure systems that drive earth's weather and climate.

3. The concentration of water vapor molecules in the atmosphere is at least 24.4 times that of carbon dioxide. When the liquid and solid-state water in the atmosphere is added, the concentration of water dwarfs the concentration of carbon dioxide. In addition, the liquid and solid-state water in the atmosphere increases the effectiveness of water in controlling the greenhouse effect to the point that carbon dioxide's role is nearly negated.

The above three facts constitute the second most important fatal flaw in the global warming theory.

Water is the most widely distributed substance on the

earth's surface. Seas cover 72% of the earth. When you add to that the surfaces of lakes, streams, wetlands and plant communities that transpire water, the figure is even higher.

When the wide distribution of water is combined with its total mass and amazing physical properties, it dwarfs the effect of all other substances in the control and modulation of earth's climate and environment.

If you want to understand the truth about global warming you must understand water's roles in the environment. Its net impact on the slight warming of the earth's surface that has occurred over the last 110 years is not equal to zero as is assumed by the global warming doctrine.

Unlike carbon dioxide, methane, ozone, nitrous oxide and other minor greenhouse gases, water should be referred to as a greenhouse substance. That is because it exists in the atmosphere as vapor (gas), water (liquid) and ice (solid). This means that water absorbs a much broader spectrum ('color' range) of sunlight and radiation given off by the earth than any other atmospheric component.

An example: According to Utah State University Climate Center data for Aug. 22, 2018, between 11:00 am and 4:00 pm at the Santaquin, Utah west site (an automated weather station), the total solar radiant energy intensity varied between 1,096 to 1,125 W/m^2 (Watts per square meter) under clear skies.

A large, very intense thunderstorm (with hail) moved into the area at 4:00 pm and the light intensity dropped to less than 17 W/m^2, equal to very late twilight levels, in 15 minutes.

This data simply describes in numbers what you know. When clouds get between you and the sun, much of the light is reflected into space, or absorbed by the water in the cloud. This is because water is extremely effective in absorbing radiant energy, whether from the sun, or from the earth's surface.

There is water vapor you can't see. There are thin clouds and dark clouds depending on how much water is between you and the sun. More water means less sunlight gets through. Rhetorical question: Have you ever "enjoyed the shade" created by the 1/25 of 1% of the atmosphere that is carbon dioxide?

The answer to the question above is obvious. Carbon dioxide is present in the atmosphere at approximately 10% of the concentration of water vapor. If the liquid and ice states are considered the level is less than 1% of the total water. It does absorb some of the sunlight moving toward earth and radiates part of that energy back into space (a cooling effect). This cooling effect is ignored by global warming advocates (See: p. 116 Appendix #3).

It also absorbs small amounts of the radiant energy given off by earth's surface, which has a warming effect. Its low concentration in the atmosphere, and limited absorption spectrum compared to the three states of water, limits its impact in both processes at levels far below the impact of water.

Chapter 8: The Role of Heat from Earth's Core in Controlling the Earth's Surface Temperature

In this chapter we will discuss the mechanisms and degree of importance of how heat from the earth's core influences earth's surface temperature. This discussion will include the contributions of volcanos, magma flows, tectonic plate movement, and heat rising by convection from magma chambers.

I realize that you can find articles which purport to debunk the idea that volcanos are a significant factor in warming of the earth's surface. This line of thought ignores several very important facts.

First: The volcanic activity that we see is limited to 25% of the earth's surface that is covered by land.

Second: The number of volcanos under the seas and their size and activity is not adequately addressed in these articles. It is known to be significant, but there is much we do not know. The earth's crust is thinner under the oceans so they may be more abundant there than on land.

Third: It is becoming apparent that heat from the earth's core, rising upward under the polar ice caps as volcanic activity and in other forms, is having a destabilizing effect on the Arctic and Antarctic ice sheets.

Fourth: In addition to volcanos, there are huge amounts of heat that move to the earth's surface from magma flows both on land and under the seas. This heat is particularly important along tectonic plate margins. Magma infusion into the tectonic plate junctions is the force behind plate movement.

Fifth: Significant amounts of heat move by convection through the earth's mantle into the seas. This heat rises to the ocean surface in warmed water.

Sixth: Interactions between heat from the earth's core and the earth's waters multiplies its impact.

The photo below shows Nishinoshima Island (Island of the East [English]) taken on Oct. 18, 2015 (courtesy of the Japanese Coast Guard). The photo shows the active lava flow emanating from the vent along with a gas plume and water discoloration. The island at that time was approximately 1.9 x 1.9 km in size at the water line.

This island is the result of a volcanic eruption that broke the surface of the Pacific in 2013 and has been active since that time, except for a few months from late 2017 to July of 2018. It is located near the junction of tectonic plates and the base is approximately 2,300 feet below the surface.

In 1973 a new island emerged from the sea in this area, and in 2013, in a spectacular eruption, this island emerged just south of the first. Small earthquakes centered in this area have been frequent since the initial 1973 eruption. In situations of this type, earthquakes are usually associated with magma movement (a heat source).

This island is one of a number to emerge from the oceans over the last 12 years from the Samoan Islands, Indonesia, the South Pacific, east of the Kamchatka Peninsula, Iceland and others.

When an event such as the emergence of an island through the sea surface is examined, there are two physical impacts that are apparent.

The first is the amount of material that is required to build the base and height of an island upward through the surface. The quantity of material is vast because as it emerges from the magma chamber it is extremely hot and tends to spread widely.

The second physical impact results from the temperature of the magma and the specific heat of what becomes the basalt that forms as it cools. The temperature of the magma will range from 1000^0C to 2500^0C and in some cases it is significantly hotter than that.

The heat carried by the magma deposited and the magma moving upward through the earth's crust is moved to the ocean's surface directly in heated water, because hot water is much lighter than cool or cold water and rapidly moves to the surface.

The resulting warmer ocean surface temperature causes increased evaporation which increases the water content of the atmosphere. This increased humidity and water content then increases the greenhouse effect, further decreasing the already minor impact of atmospheric carbon dioxide.

The above mentions only those volcanoes that have risen above the sea's surface. With over 72% of the earth's surface being covered by oceans it would be expected that the numbers of active volcanos and lava flows that we don't see would be significant.

To assist in understanding this topic as it relates to earth's surface temperature, let us consider an example of what happens below the sea surface and usually goes unnoticed.

In 2012, a huge block of floating pumice rock was spotted by passengers on a flight over the Pacific Ocean. A 2015 paper by volcanologist Rebecca Carey from the University of Tasmania and Adam Soule (WHOI) reported on an examination of the site of the underwater volcano known as the Havre Seamount.

The authors reported on what they found, 4,000 feet below the surface: "The caldera, which spans nearly 4.5 kilometers (3 miles) discharged lava from some 14 vents in a massive rupture of the volcanic edifice, producing not just pumice rock, but ash, lava domes, and seafloor lava flows".

Quoting Dr. Carey further: "It may have been thankfully buried under an ocean of water, but for a sense of scale, think roughly 1.5 times larger than the 1980 eruption of Mt. Saint Helens, or 10 times the size of the 2010 Eyjafjallaokull eruption in Iceland".

There are multiple ways in which the vast quantities of heat from the earth's core moves to the earth's surface. It requires a multidimensional assessment of these forces to gain a true picture of the influence of this heat source.

It cannot be done with a unidimensional analysis, ie by examining only the visible volcanic activity and drawing conclusions from that small fraction of the total impacts.

The high-resolution seafloor topography above, from the 2015 expedition, shows significant heat sources from the peak in the upper center and from lava flows along the rim and in the bottom of the crater. This is 3

years after the initial eruption. The crater shown is 2½ times the size of the volcanic island Nishinoshima shown earlier.

Author's note: In the book: Fatal Fails: Scientific Problems with Global Warming, you will find other, highly significant examples of how volcanic activity and other sources of heat from the Earth's core has contributed to the rise in the Earth's surface temperature during the last 110 years. Also, the impact of heat from the Earth's core on the stability of the Arctic and Antarctic ice sheets is a significant topic in that work.

Additional note: The reader may find it interesting to visit YouTube and view some of the videos available of under-water volcanic and magma flow activity.

In the most remote areas of the world the ubiquitous distribution of aerosols can be seen. Note in the distance the fuzzy nature of what you see. The fuzziness and color shift is the result of light scattering by aerosols in the atmosphere.

Chapter 9: The Role of Aerosols in the Control of Earth's Surface Temperature

Quoting: https://earthobservatory.nasa.gov/features/Aerosols

"Take a deep breath. Even if the air looks clear, it's nearly certain that you'll inhale tens of millions of solid particles and liquid droplets. These ubiquitous specks of matter are known as aerosols, and they can be found in the air over oceans, deserts, mountains, forests, ice and every ecosystem in between."

"They drift in Earth's atmosphere from the stratosphere to the surface and range in size from a few nanometers–less than the width of the smallest viruses–to several tens of micrometers–about the diameter of human hair. Despite their small size, they have major impacts on our climate and our health".

Quoting Further: "The bulk of aerosols–about 90 percent by mass–have natural origins. Volcanoes, for example, eject huge columns of ash into the air, as well as sulfur dioxide and other gasses."

An example of natural aerosols are the phenol and terpene materials given off by forest trees which cause the late day "Smokey" appearance in the Smokey Mountains and the blue haze that gives the "Blue Mountains" name to several forest areas in the western United States. They are also the compounds that give the forest that fresh scent which is so refreshing.

Human caused aerosols constitute approximately 10% of the total world-wide. NASA appears to include some classes of particulates, as having natural origins, that are increasingly (over the last 150 years) becoming more directly related to human activities.

Two examples are, aerosols arising from forest fires (75-90% of which are human caused) and wind-blown dust from deserts and other areas. These areas are increasingly used and the surface disturbed, by human activities giving rise to dust. For these reasons the 10% figure is probably low.

Aerosols contribute to the cooling of earth's atmosphere as they intercept, scatter and re-radiate back into space some of the sun's energy as it moves toward the earth's surface. Global warming/ climate change doctrine considers this cooling effect.

What the doctrine ignores is that these aerosols also contribute to the greenhouse effect of the atmosphere by intercepting and absorbing radiant energy given off by the earth's surface. This is a warming effect of great significance.

The impact of aerosols on the warming of earth's surface is attributed, by default, to carbon dioxide forcing of earth's surface temperature, a fatal flaw of the global warming theory.

You have now been introduced to the 4 main forces that modulate and control the earth's surface temperature. You are now equipped to understand how the greenhouse effect of carbon dioxide has been grossly exaggerated.

Chapter 10: The Role Wildfires Have Played in Increasing Atmospheric Carbon Dioxide

In recent years the average size of wildfires in the US, Canada, and several other major atmospheric carbon dioxide contributors has been increasing. This source has become a significant factor in the annual rise of the earth's atmospheric carbon dioxide content.

Politicians, global warming/climate change disciples and others have made the claim that the size and frequency of wildfires is the result of what they claim is an approximate 1^0 Celsius rise of the earth's surface temperature. This claim is not supported by any factual foundation.

Beginning in the early and mid-1990s the size of the average wildfire in the US, Canada and Australia began to increase. At that point the global mean temperature anomaly was 0.37^0C (average of 1995 and 1996). Since that time the average size of the 61,375 (annual average) human caused wildfires in the US has increased from 43 to 119 acres. The size of the 9,946 (annual average) naturally caused fires has increased to 382 acres.

During the 10 years from 2009 to 2018 there was a total of 113 million acres, or 176,440 square miles, burned by wildfires and controlled burns in the US. The number of pounds of carbon dioxide that has been, and will be, added to the earth's atmosphere because of the US wildfires and controlled burns over the 2009 to 2018 period is equal to 26 trillion pounds (over 13 billion tons). World-wide, wildfires contribute approximately 1/3 of the annual net increase in the earth's atmospheric carbon dioxide concentration.

Why have large fires became so frequent and violent? The blame rests squarely on changes that have occurred in forest management and fire control doctrine over the last 50 years and particularly since 1995-96.

Beginning in the 1970s there was a steady increase in pressure from environmentalists to change long standing forest management and fire control practices.

The pattern for the above policy changes was the 32 million acres of roadless forest lands (within the US and much more world-wide) that are regulated by the IUCN (International Union for Conservation of Nature). These regulations played a major role in the 1988 Yellowstone Fire disaster which consumed over 800,000 acres (1250 sq. mi) of forest in and around Yellowstone National Park.

A significant portion of the problems encountered in the last 30 years are traceable to the concepts of this agreement being incorporated into the general wildfire control doctrine and procedures.

Prior to 1990 the US fire control policy was to locate and extinguish all fires as quickly as possible. That policy was replaced by a "more natural," let it burn if it was naturally caused policy. This delays the start of suppression efforts in many cases until the fire is determined to be human caused. A delay of minutes can change an easily controlled fire into a disaster.

Changes, such as restrictions on timber sales to harvest dead and diseased trees for building materials, fuel load reduction practices, and other prudent forest management tools were made with disastrous results.

These policy changes are responsible for the average size of wildfires increasing from 43 to 119 acres between 1990 and 2019. A huge increase in wildfires of 10,000 to 100,000 acres and larger occurred in this time period. This led to the average annual acres consumed by wildfires (in the USA) increasing by nearly 300%.

An example of this is what happened in the forests of Idaho and Utah during the drought years of 1999 to 2006. During this time period the drought stressed trees, especially Engelmann Spruce and Douglas Fir, which use a lot of water and are drought sensitive, died across several million acres.

The water stressed trees were very susceptible to bark beetles which devoured the cambium and phloem, weakening and killing the trees.

This resulted in tens of millions of dead, standing, mature trees mixed with young developing trees of more drought tolerant species. Attempts were made by forest managers in many areas to allow the harvest of these trees through timber sale contracts.

The 2005 forest management act attempted to make it easier to remove these trees, but those timber sales were adamantly opposed by environmentalists.

The result was that those trees, enough to build millions of homes, still stand, rotting or burning.

Many areas with this type of forests have experienced significant numbers of large and very large wildfires. This same type of extreme fuel loads accumulated in many public-land forests from California to the Carolina's and from Florida to Alaska.

Overt neglect and forced ignoring of fuel load buildup played a major role in many huge, destructive wildfires. For example, the Happy Camp Fire Complex in California during the 2018 fire season began in, and burned through, an area which had not experienced any major fuel reduction practices in 126 years.

The area literally exploded in a firestorm which destroyed most of the city of Paradise CA and killed 84 people (The devastation continues; As of 9/10/2021 over six million acres of forest has been destroyed in California in the last 3 years). The destruction continues as huge fires rage in California and Oregon each year.

It is imperative that proactive forest management fire control practices be implemented immediately. We know how to do it (it was done for 70 years prior to the change in control procedures which occurred during the 1980s and 90s). The average fire today burns three times the area consumed by the average fire just 25 years ago. If this is not done we will see continued increases in wildfire destruction during dry periods.

The photo below, taken 9/25/2010 shows a mountainside in the Crandall Canyon area of central Utah (USA). This area burned in 1934 and had partially recovered during the intervening 85 years.

Growing under the beautiful aspen forest was a developing forest of spruce and fir trees which would have replaced the aspen in the next 2 or 3 decades. The area shown constitutes roughly 400 acres. Each acre at this stage of development will sequester roughly 25,000 pounds of carbon dioxide per year. This figure includes the carbon dioxide sequestered in tops and root zones of the trees.

The following photo shows a small area (about one acre at the top center of the orange aspen area in the above photo). This photo was taken on October 16, 2019, just 15 months after a wildfire destroyed this forest.

The general area that this 19,000 acre fire burned has mostly north and high elevation slopes. This photo is typical of those areas.

The area shown is much smaller than in the pre-fire photo, but you should get a feel for how many trees were destroyed by multiplying what you see by the number of acres (approximately 19,000 acres were burned). This was not a large fire by current standards, but it was very destructive.

I began to hunt deer in this canyon in the late 1950s. My father cut timber here for 10 years to support his siblings beginning at the age of 12 in 1922. He sold the logs to local coal mines for use as mine props.

Let's examine the environmental tragedy which results from a wildfire which is never fully recognized.

When a wildfire occurs, the first thing that happens is that the small branches, bark, needles, leaves and organic materials on the forest floor go up in flames. That material is oxidized, becoming carbon dioxide, water and ash very quickly.

If the forest is dry and dense, with a buildup of fuel from overgrowth and brush, more of the heavy wood will burn during the fire. This fire burned in early June when the trees were not as dry as they would be in July or August, so less of the sequestered carbon dioxide in the large diameter wood was released during the fire. As you can readily observe from the photo, when the fire was extinguished there were many partially burned trees left.

Author's note: Excuse me for using a lot of numbers in the next few pages. It is necessary to help fully understand the magnitude of the impact of forest fires on the carbon dioxide concentration of earth's atmosphere.

The above ground portion of the trees in this instance represents about 620 trees per acre which would average approximately 225 pounds (dry weight) per tree. That is 139,500 pounds of wood per acre (69.75 tons) which will eventually decay, releasing the carbon in the tree structure into the atmosphere as carbon dioxide.

Because of the chemical composition of wood, when it burns, or decays, considerable oxygen is utilized. This addition of oxygen increases the weight of carbon dioxide produced by approximately 34%.

The total carbon dioxide released (from the above ground portion of the forest) during the fire and decomposition phases will be equal to 1,775,835 tons.

The below ground portion (root systems) of the forest will generally be equal to 60 to 80% of the sequestered carbon content of the tree tops. The soil environment will chemically sequester a sizable portion of the carbon dioxide released as the roots decay so I will not estimate that portion of the equation.

That leaves only one more adverse impact of a fire on carbon dioxide sequestration. This one is usually the most important.

When a forest matures, the canopy reaches a maximum extent, as does the photosynthetic potential. The height of the trees continues to increase as does the diameter. At this point, the percentage of the product of the photosynthetic process which is deposited in the woody portions of the trees (trunk, limbs, and root system) also reaches a maximum.

In short, the mature forest sequestration of carbon dioxide reaches a very high level, which can be maintained and added to for long periods, even hundreds of years. In a fire ravaged area, as is illustrated by the preceding photo, it will take 30 to 50 years before a complete tree canopy is re-established.

During this phase of forest regrowth an average of 60% to 70% of the carbon dioxide that would have been sequestered will remain in the atmosphere.

In some environments, such as the West slopes of the Cascade Mountains in North America, a mature forest may be regrown in 25 to 30 years. In dryer areas such as the Rocky Mountains, it will take at least twice that length of time, or more.

Based on what is written in the preceding two paragraphs, if we assume that it will require 30 years to develop a mature canopy, the following will happen across the 19,000 acres shown and described above.

Over the next 30 years, only approximately 40% to 50% as much carbon dioxide will be sequestered as would have happened had the fire not occurred. That would amount to 4,631,500 tons of carbon dioxide which will not be removed from the atmosphere and sequestered.

When this total and the total carbon dioxide released by the fire and subsequent decomposition of the burned trees are combined, 6,213,250 tons of carbon dioxide will be left in the atmosphere.

During the 2016 to 2018 fire seasons, over 13 million acres were destroyed by wildfires and controlled burns annually. The 19,000 acres involved in the fire discussed herein constituted 0.146% of the annual average acres burned by wildfires between 2016 and 2018 within the USA (48 adjacent states + Alaska).

Extrapolation of the above estimated carbon dioxide that will remain in the atmosphere means that 4,248,800,000 tons of carbon dioxide will not be sequestered annually. That is enough carbon dioxide to increase the earth's atmospheric content by roughly 0.9 ppm, or 30% of the annual net increase of 2.64 ppm each year for the last three years.

Author's note; I offer the above extrapolation with the reservation that it is probably low, because as the carbon dioxide level in the atmosphere increases (as it currently is) the rate of photosynthesis in most plants also increases. That would result in the number of tons of carbon dioxide left in the atmosphere being larger.

It should be remembered that this extrapolation takes into consideration only the land area of the 48 adjacent states and Alaska (the continental US) and involves three fire seasons. That is approximately 5% of the total land area of earth. These facts mean that the impact of wildfires world-wide is a highly significant contributor to the rising carbon dioxide content of earth's atmosphere.

If the acres of forests lost to wildfires and urbanization are not decreased, the annual lost sequestration will continue to rise. This could soon be equal to, or greater than the net carbon dioxide released into the atmosphere by other human activities world-wide.

For a complete discussion of the issues addressed in this chapter please see: Fatal Fails: Scientific Problems with Global Warming.

Author's note: Few of us really see the destruction done by wildfires. We see short video clips of the fires on TV, and we may drive past a fire after it is controlled at 60 to 80 mph, but one can't fully comprehend what they do unless we experience being driven from our homes or hike through and study a recent burn site.

During the summer months of 2019, the photosynthetic use of carbon dioxide from the atmosphere, along with the other processes that remove this gas, resulted in a significant decrease in its concentration.

On 5/15/2019 the carbon dioxide level in the atmosphere was 415.7 ppm. In July the value was 411.77 and by September 3, the level had dropped to

408.63 parts per million. It remained close to that value through October, despite the record breaking contributions of wildfires in California and Australia.

This illustrates how fast the photosynthetic and other sequestration processes can remove carbon dioxide from the atmosphere.

During this 5-month period, the atmospheric concentration of carbon dioxide was reduced by 7.07 parts per million (415.7 ppm minus 408.63 ppm). This equals 40.12 billion tons of carbon dioxide which was removed from the atmosphere.

Earth's atmospheric carbon dioxide content dropped by about 1.4 ppm per month during this 5-month period. Such is the power of the photosynthetic and chemical sequestration processes.

The above paragraphs suggest a solution to the rising carbon dioxide content of the atmosphere (to the extent that it requires a solution). In short that solution consists of the following:

1. Decrease the loss of forests by wildfires worldwide to pre-1990 levels;

2. Reforest at least 1 billion hectares of earth's surface.

3. Manage large swaths of earth's forests to assure maximum levels of carbon dioxide sequestration.

This could be done economically, in a way which will be self-sustaining after 15 to 20 years. It would require a total investment of approximately 1-2% of what the "New Green Deal" is estimated to cost if that course is adopted, and it will solve the problem, not make it worse (For a discussion of how the proposed Green New Deal will make Earth's environmental problems worse; see the book Fatal Fails).

During the time between Sept. 2017 to Sept. 2019, the level of carbon dioxide in the atmosphere rose by 5.28 parts per million (2.64 ppm/yr). The 11-year period from 2009 to 2019 averaged slightly under 2 parts per million per year.

To end this discussion in this source, and to set the stage for the above-mentioned discussion on this subject, let us consider several rhetorical questions.

Rhetorical question 1: Given that the 2017-2020 fire seasons in the US, Central Canada, the Amazon and Australia were the most destructive in history, is it surprising that there was a sudden increase in carbon dioxide in the air?

Rhetorical question 2: What portion of the 2+ ppm per year rise in the atmospheric carbon dioxide concentration would disappear if there was a return to the more aggressive wildfire control practices used before 1990?

Rhetorical question 3: How much of the carbon dioxide increase in earth's atmosphere has resulted from the adoption of the IUCN regulations throughout much of the world?

Rhetorical question 4: Is it a good practice to allow agenda driven, emotion-based changes in policies and procedures to take place without first examining what the unintended consequences will be?

Rhetorical question 5: If we could aggressively limit wildfires world-wide what would be the impact on atmospheric carbon dioxide content?

Rhetorical question 6: Given that much of the earth's most severe poverty haunts the peoples of the tropics, could you contemplate what an effective reforestation program in those areas could mean?

As you contemplate the above questions consider that: Well managed forests moderate (cool) the environment, provide huge economic benefits, can provide conditions favorable to tropical food production, build soil and stabilize and purify water supplies.

Author's note: Chapters 10 and 11 are included in this work to help the reader understand that the environmental problems this earth's population faces are much more complex than the disciples of the global warming theology (theory) comprehend. These two chapters will help you understand that there are effective alternative solutions to these environmental challenges.

Chapter 11: Reforestation: A Major Component of The Solution to Many Environmental Challenges

The facts presented in Chapter 10 suggest a rather obvious series of actions that could be taken to stabilize or even reduce the atmospheric carbon dioxide concentration. In this chapter I will briefly lay out the general principles of this critical subject.

The loss of approximately one billion hectares (2.47 billion acres) of forest lands since 1920 world-wide has been responsible for a significant portion of the rise in the atmospheric carbon dioxide content. Since 1700 a total of 5.5 billion hectares (13.59 billion acres) of forest has been lost world-wide.

If we had as much forest now as in 1920 the concentration of carbon dioxide in the earth's atmosphere would reach an equilibrium state somewhere between 360 to 390 ppm (equilibrium state is the level at which the increased photosynthetic and chemical sequestration rate is equal to the rate of carbon dioxide release on an annual basis).

The highly publicized figure of 40 billion tons of carbon dioxide added to the earth's atmosphere each year represents the total human caused releases. The photosynthetic and chemical sequestration processes discussed earlier reduce the net increase to 11.35 billion tons.

Forests are of extreme importance to earth's environment. Consider the following facts:

1. Forests scrub from the atmosphere many different pollutants and greenhouse gases.

2. Forests assist in mitigation of variations in water supplies and quality. Both supply and quality are increasingly important as the population of the earth expands. Because of the loss of approximately 50% of earth's forest areas since 1700, climates have changed and potable water supplies have decreased.

3. Forests cool the environment, an important quality of life factor in the tropics.

4. Forests build and restore depleted soils and increase available mineral nutrients.

5. In the tropics, use of the right species, in properly designed forests can provide a conducive environment for the production of food crops. Peppers, sweet potatoes, beans, tomatoes, several types of spinach and a significant number of tropical fruits such as passion fruit and pineapple thrive in the partial shade of open canopy species of tropical trees.

6. Forests yield huge economic benefits for those who live and work in and around them. These facts are particularly true in tropical areas. These economic benefits are well documented in the 2012 edition of "The State of the World's Forests" report of the Food and Agriculture Organization of the United Nations.

We are told that there is only one way that we (the human species) have a chance at survival. All must sacrifice our living standard and freedom to make many choices that affect our wellbeing on the Altars of the secular religion of global warming.

The religion's prophets and disciples dictate that all

of this must happen, or we face certain extinction. The future ruling class travels widely for conferences, to communicate the dogma and for recreation while ensconced in prestigious, well paid positions that support their lavish lifestyles. Lifestyles which must and will be maintained for those who, in their view, are the solution to the problem.

The above is a dire yet accurate prediction of what will happen unless the global warming doctrine is resisted.

Rhetorical question 1: Is it true that we are doomed if we don't give up most or all our fossil fuel use very soon? Rhetorical question 2: Is it true that there is only one solution? I would submit that the answer to these two questions is an emphatic no.

Even if rising carbon dioxide levels were the problem, there are solutions which do not involve drastic changes in the lives of the 7.8 billion people who live on this earth. There are affordable, steps which may be taken in a multi-component approach to slow and even reverse the rise in atmospheric carbon dioxide while enriching the lives of earth's inhabitants.

The remainder of this chapter enumerates the most important components we need to consider in an effort to improve the environment, address poverty, and slow or stop the accumulation of carbon dioxide in the atmosphere. These components are:

First: There is a dire need to reduce the numbers and average size of wildfires, not only in the US but world-wide.

Second: Conversion of forest lands into roads, parking lots and other practices which remove or reduce the plant cover needs to be limited. In the alternative, these areas must be replaced by intense reforestation efforts in other areas.

Third: A worldwide effort at reforestation needs to be undertaken and financially supported. A system of private and public investment funding and/or tax incentives could do a lot to replace the one billion hectares of forests lost since 1920.

Fourth: The exemptions given to countries in the India, China, South East Asia area for the last 30 years, that allowed them to industrialize with little or no attention to emissions control, need to be lifted.

Fifth: Tropical agricultural methods need to be altered to increase the planting of long-lived trees for selective harvest and to create a more favorable environment for diverse food production.

Sixth: Fertilization of selected areas of the southern hemisphere seas needs to be researched, funded and incentivized to make those vast areas more productive. These areas could produce vast quantities of fish and other seafood while sequestering additional quantities of carbon dioxide.

The above efforts, incentivized and managed could help solve many of the world's environmental issues, while enriching the lives of all inhabitants of this earth.

Chapter 12: A final Critical Look at The Global Warming Numbers and Methods

What They Really Tell Us

The Freshly Pressed, Ice Cold, Apple Cider Allegory:

A home-owner who was into gardening and raising apples came to harvest time in the fall with an abundance of fruit on his trees. He was an advocate of "Organic Gardening" who had not began to use mating disruption, so most of the fruit had at least one worm hole in it. He picked the fruit and as he filled the 4 baskets, he separated the best fruit into one basket (those with only one worm hole), the wormy fruit (those with 2 or more worm holes) into another basket, the blemished fruit into another, and the soft fruit into a 4th. When he finished he looked at the baskets of fruit which were heaped with apples and considered each.

Basket 1: Contained the best fruit. He was proud of it, large, crisp, juicy excellent fruit (except for the worm holes).

Basket 2: This fruit was fruit that had more than 1 worm hole in them. He knew that all the worms had matured, left the fruit, found a place to pupate over the winter and complete their life cycle the next season. All that was left of those pesky things were the holes, and yes the apple tissue that had passed through the worms gut. He had heard a professional entomologist refer to this as "frass," a nice scholarly term, much better than worm poop.

Basket 3: This fruit was very nice but had one or more blemishes on them. They were just right for cider, after all the blemishes were only skin deep.

And the worms loved them, as indicated by the number of worm holes.

Basket 4: This fruit was large, had been crisp, was very juicy, just that it was a little like the apples you sometimes buy in August that have been in storage for 12 months. It was not rotting, but soon would be if he did not use it. Oh, and it had worm holes in almost every apple.

He arranged with a friend to borrow a cider press in 2 weeks to juice his apples, and not having room elsewhere, he sat the 4 baskets under a large shade tree. He watched the baskets closely but noticed each evening a large dog and 2 smaller dogs pass the baskets, lift a rear leg next to baskets 1 and 4 and mark their territory. He thought, I should stop that, but after a long days work I deserve to relax with my coffee and news program.

On the day he had chosen to press the apples he had the thought: I really want 10 gallons of apple cider, and I know that each basket will yield about 2.5 gallons. If I use all the fruit I can get the 10 gallons, and besides there are no worms in the fruit with worm holes in them, only the frass and possibly a little mold in the worm holes. The soft fruit seemed to be very juicy, and the blemishes were only an appearance issue.

Each bushel weighed almost 50 pounds and after the dog urine had evaporated there could not be more than a few grams of residue. So, what is a few grams out of 90,800 grams of total weight (he was conversant with the metric system).

So, he juiced all 4 baskets and ended up with 10 gallons of cider. Oh, congratulations, he has invited you and your significant other over for a night of cards and fresh pressed, cold, apple cider and doughnuts.

I'll let you consider what this has to do with, anomalies, temperature spreading, removing "mistakes" from raw weather data, and homogenization of climatic data sets. You are not invited over for fresh, cold apple cider, but you are invited to read on and consider what you read. (You may want to review Appendix #1 {p. 110} that deals with "Normalization of Deviance and GroupThink and keep it handy for quick reference. See if you can recognize any of Vaughn's 8 symptoms, or the ones I added at the end.)

Earlier, at the end of Chapter 2, I wrote: The bottom line is that the true increase in the earth's surface temperature lies somewhere between 0.45^0 and 0.55^0C, not the 1.1^0C that has been recently claimed (NASA's 2019 numbers range from 0.9^0C to 1.1^0C).

We also reviewed how the underlying assumptions of the global warming theory disregard the four main forces that have contributed to the warming of the earth's surface over the last 110 years.

In short, 100% of the causation of the relatively small increase in the earth's surface temperature was not carbon dioxide forcing. The four main environmental forces that control and modulate earth's weather and climate are the sun, water, heat from the earth's core and aerosols.

As demonstrated by the allegory of "My Strict Diet," correlation does not prove causation.

In this chapter I will provide additional information

that indicates the numbers provided by global warming advocates are suspect because they are routinely edited, changed, manipulated and otherwise fudged to obtain doctrinally friendly results.

The changes keep occurring, and each major step increases the cooling bias correction; meaning, they increase the magnitude of the claimed warming.

This will be a brief synopsis of the high points of this topic. For those who would question my conclusions or desire additional information I would refer you to "Fatal Fails: Scientific Problems with Global Warming".

There are some key facts that one must know in order to understand what has been happening, particularly during the last 30 years.

First: I need you to understand that the raw data (what the thermometer records) is, in reality, being changed.

Author's note: In several places in this work I write that the raw data is being changed. In a technical sense that is not correct. The raw data (According to NASA) is not being changed, it is just not being used and it is very difficult to find and interpret (A demonstration of twisted logic?). What is being used, is the "unadjusted data": the raw data after it has gone through a 5-step series of "corrections". Is the use of the term "unadjusted data" an additional twist to the above twisted logic? There have been several versions of "unadjusted data" the GHCNv1, GHCNv2, GHCNv3, GHCNv3.1, GHCNv3.2 and as of May 2019 the GHCNv4.

Second: Before 2011, there were data changes but they were not as extensive as after 2011, when the GHCNv3 dataset was adopted by NASA and other

Global Warming advocates. During the first 100 years of the "global warming period" the anomaly increased very little. According to the NASA web archive the anomaly was equal to 0.07^{0}C in 1978.

Third: When you see a graph, data or other publication originating after 2011 you must consider that there have been extensive changes in the data. The "unadjusted data" (what is used) has undergone the before mentioned 5 step "correction" and "homogenization" process.

Even the 4-Agency graph that I use several times in this work is based on the GHCNv3 dataset which flattens the graph between 1998 and 2011. Before that, the raw data showed a significant dip during this period (equal to 0.3^{0}C).

The dip referred to in the last sentence of the above paragraph was the dip referred to during the "climate gate" fiasco (and shown by the raw data [red line] in the Berkley Earth graph which appears in this work).

To understand what is happening, we need to visit the occurrences that became known as "Climate Gate". I find this necessity distasteful, but not as much as I found that series of events astounding and repulsive, even sickening, because of what it says about what "science" has become.

A hacker obtained access to a server at the Climate Research Unit at the University of East Anglia (UEA) (circa 2009) and "hacked" many emails and other documents which were then published.

This led to a heated controversy between the "disciples" and "deniers" of global warming on a number of different things written and discussed in those documents. One of the focal points of this controversy was a question asked by one of the authors of the emails concerning the fact that the temperature trend was going down, instead of up, like the carbon dioxide concentration trend (the correlation was negative).

The question was, "How do we hide the dip?" The answer was, "Use the Mann trick". This had reference to a methodology employed by the author of the hockey stick theory.

In simple terms, the technique is to use homogenization of a different data set (or portion of one) with the data set the researcher has been using, to get a positive outcome (an outcome that is doctrinally friendly or more striking i.e. an anomaly increase).

The claim was that the use of the word "trick" was taken out of context and that the technique was a recognized procedure often used in climate science research.

There did not appear to be a denial that the technique changed or accentuated the result. Rhetorical question: If a procedure or technique is used for the purpose of changing the results of an experiment is it "a trick" or is it "science"?

A panel of Dr. Mann's peers was called together to consider if he had "seriously deviated from accepted practices within the academic community for proposing, conducting, or reporting research or other scholarly activities”. The panel concluded that he had not. The disciples of global warming were elated and claimed vindication. I have twice offered in this work examples of the results of the homogenization procedure.

Rhetorical question: Does the fact that this practice (homogenization of data sets to obtain a desired result) “does not seriously deviate(d) from accepted practices within the academic community” serve as vindication, or a much broader indictment?

The above question is extremely important because, as you will see in this chapter, the changing of data, the homogenization of data sets, use of rolling multi-year averages (LOWESS smoothing) and other techniques, often change the results and/or unjustifiably accentuate trends.

Again, I ask: Does the fact that these practices do not “seriously deviate from accepted practices within the academic community for proposing, conducting, or reporting research or other scholarly activities” give you confidence in the results?

True science is not a “everybody does it, so it is ok” proposition. At least it is not according to my scientific training.

This methodology is, in my opinion, a totally non-scientific, even anti-scientific procedure. Simply stated, this trick allows a researcher to: When a researcher fails to obtain the result he wants, as he “runs his model” he uses a different data set, hybrids it with his own and re-runs the model. If he chooses the right data set, and their vagrancies, deficiencies and tendencies are well known, the non-significant result is "presto" turned into a significant one.

May I observe that using a method such as this is much easier than employing science, you do not run the risk of a result that does not support the global warming doctrine. Also, you do not risk attaining the title of "heretic" with the attendant label of "denier" which may bring an end to your career and standing in the "scientific community."

What happened next? In May 2019 a new version of the GHCNv3 data set was adopted. It is the GHCNv4 data set, and again, as in 2011, with the adoption of the GHCNv3 data set, it reversed a "short term" downward trend of the global mean surface temperature anomaly.

It will be argued that a 2-year downward trend (as happened in 2017 and 2018) or a 14-year downward trend as occurred from 1998 to 2011, were weather, not climatic trends, because they were not of enough length to establish a climatology.

A climatology is defined as a 30-year trend. Obviously, the frequent adoption of different data sets can prevent any trend, that is reversed by the new data set, from becoming a climatology. Global warming advocates require a 30 year trend to qualify as climate change.

Since 2011, essentially all of the global warming advocates (including NASA) have used the GHCN data set series. There was GHCNv1, GHCNv2 prior to

2011, then in that year GHCNv3 was adopted. There were two modifications of that set, GHCNv3.1 and GHCNv3.2 and in May 2019 GHCNv4 was adopted and is currently the holy grail of datasets.

Each of these datasets contain additional data corrections and changes from the previous. Each of these dataset versions also increased the numbers and/or magnitude of cooling bias corrections (meaning they increased the magnitude of the temperature anomaly).

The GHCNv3 and GHCNv4 versions also add significant numbers of additional observation sites around the world. The above mentioned vindication of the "Trick" occurred in 2011 and appears to have been immediately adopted (as license to change data) by the authors and advocates of the GHCN data set series.

The following is from: "GISS Surface Temperature Analysis: The Elusive Absolute Surface Air Temp. "https://data.giss.nasa.gov/gistemp/faq/abs_temp.html" Updated 2019-02-12:21:26

This document is one that has been around for years (updated periodically) and consists of a series of questions which are meant to help one understand what the GISS Surface Temperature Analysis System is and why it is used.

There are several internal inconsistencies contained in the document, but I will not deal with all of them in this work. I would refer those who would be interested in a full discussion to "Fatal Fails" which is more comprehensive.

It is important to note the date of the last update; Feb. 12, 2019. This date is just 3 months before NASA adopted the GHCNv4 version of the GHCN data set series, in May 2019. I would further call your attention to the last question and answer contained in the Feb. 2019 update which reads: Q. What do I do if I need absolute SATs (Surface Air Temps), not anomalies?

The last sentence of the answer to the above question reads: “For the global mean, the most trusted models produce a value of roughly 14^{0}C, ie 57.2^{0}F, but it may easily be anywhere between 56^0 and 58^{0}F”.

The admission here is that the GISS anomaly system has an error of at least ±1^{0}F on a global basis. On a regional or local level, the error is even greater. What does this error mean? Briefly it means that any actual anomaly of less than 1.0^{0}F (0.6^{0}C) cannot be judged to be different from the mean (reference period or normal). It is said to be within the margin of error.

I apologize for the following “trip into the technical weeds”. It is necessary. If you will work a little you will be able to realize what a house of cards this scheme has become.

Examine the graph of the Annual Temperature Anomaly on p. 92. Note that the last year represented on the graph is 2014 and that all four of the indicated agencies estimate of the global mean anomaly ranged between 0.6^{0}C and 0.7^{0}C for that year. Also, note that during 1880 to 1910 the annual temperature anomaly decreased.

Remember the Feb. 2019 NASA document mentioned previously describing the GISSTemp SAT system.

I would like to illustrate the importance of the admission that there is a 0.6⁰C error in the system.The graph below is from NASA and shows the global mean temperature anomaly for the years of 1880 to 2014. You have seen this graph before.

Note the 0 line on the graph. This is the 1951 to 1980 average mean global temperature or the "normal" surface air temperature.

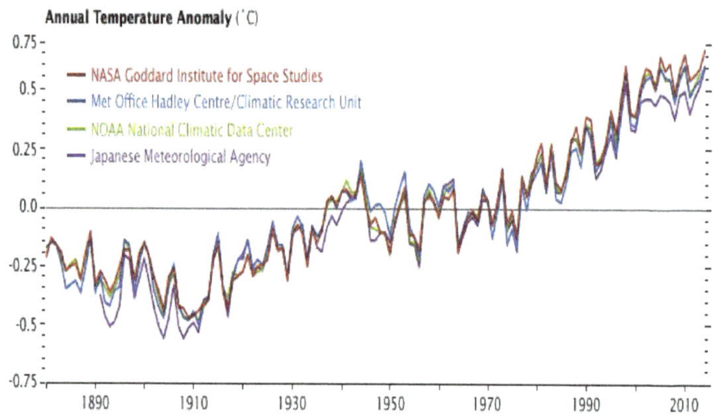

The first graph below shows the Global Mean Estimates based on land data only (through 2018).

Notice the positions of the 2 red lines at, -0.6⁰C and +0.6⁰C on the two graphs below. What the statement which reads, "but it may easily be anywhere between 56 and 58⁰F" means is, any position between these 2 red lines could, in fact, be the mean global average temperature for the 30-year "normal" or reference period.

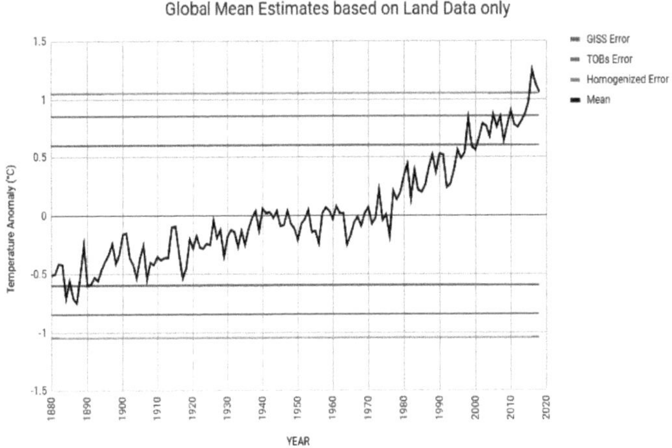

So, what if the true global average temperature for the 30-year "normal period" was at or near the lower limit of the admitted error range? If that were the case the graph below presents the real picture.

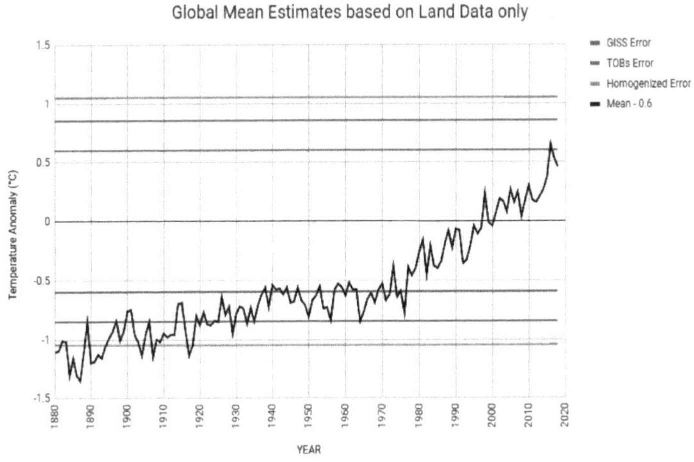

Remember that the 0 line represents the estimated

global mean temperature for the period of 1951 to 1980 (57.2^{0}F). If the estimate was 1^{0}F higher than the actual global mean temperature, the true anomaly did not exceed the admitted error limits until 2016, then dropped below the upper error limit for 2017 and 2018.

You will note that there are blue lines at 0.85^{0}C and -0.85^{0}C (equals a net change over the real data of ±0.25^{0}C). These lines indicate the portion of the error limits "removed" by the corrections to the raw data.

You will also note that there are green lines at 1.05^{0}C and -1.05^{0}C (equals an additional ±0.2^{0}C of change over the raw data). The difference between the blue lines and the green lines indicates the portion of the true error limits "removed" by the homogenization of several different data sets to obtain a more doctrinally favorable result.

Changing the data has always been considered abhorrent in scientific circles and that should never change. If a "researcher" can change the numerical value of his observations, all he needs to conduct research is a pencil and piece of paper. This is not science, it is much closer to witchcraft!

You now have a good view of another aspect of the "settled science" behind the global warming/climate change doctrine. This is one reason I use the term doctrine, to distinguish the theory and methods from the facts.

The admitted error in the GISS system average surface temperature is ±1^{0}F (0.6^{0}C). If the data manipulation changes and dataset homogenizations are considered, that figure increases to ±1.8^{0}F (±1.0^{0}C).

If you are interested in local or regional anomalies you must be aware that the errors are even worse.

Two additional ignored error factors should be noted. The first is an altitude bias which results from the fact that the temperatures used in this system are those from weather-stations, most of which are close to, or in, cities or towns.

Most cities and towns lie in the valleys or low lands. This leaves the higher elevations under represented, not represented, or because of the temperature spreading that occurs in this scheme, misrepresented.

These areas represent a significant portion of the earth's land masses. They tend to be significantly colder than the lower areas, often by 10^0 to 20^{0}F or more. These high-altitude areas also cool down faster in the evening and warm up much slower during the morning hours.

This adds another dimension to the error because the daily average temperature (the high temperature plus the low temperature divided by 2) is used in this system. That means that the rapid cool down and slow warm up each day is very different from that of lower elevations, yielding a distorted daily average.

The second of the two errors mentioned above results from the fact that there are still significant areas of the earth's surface which are under-represented or not represented at all by actual data.

These areas' temperatures are estimated by temperature spreading and other editing techniques (including extrapolations {a scientific term which is equal to guesses}). The bulk of these under represented areas lie in the Arctic, Antarctic, South Atlantic, South Pacific, and Siberia, which are all cold areas.

When temperature estimates for these areas are made by spreading or extrapolating temperatures from near-by areas (usually hundreds of miles away), the resulting errors are large, and a great place for bias to creep (or flood), into the result.

In 2011, the use of the GHCNv3 data set began. This data set included all the previous adjustments made in the "unadjusted data". There was also inclusion of data from a significant number of additional weather stations.

This practice is highly problematic because it changes the average current temperature data relationship with the normal value (the global mean temperature for the 1951 to 1980 period). This resulted in a data set with a vastly increased upward slope between 2011 and 2016. Do you believe in coincidences?

The error introduced by the combination of changing the raw data into the "unadjusted data" and the homogenization of data sets adds at least an additional 0.45^{0}C error. That increases the total error to ±1.05^{0}C (The admitted 0.6^{0}C plus the fudge factor, of 0.45^{0}C). This extends the confidence interval beyond even the most outrageous estimates of the global surface temperature anomaly increase.

This illustrates why, once science is abandoned, as it has been with the global warming advocates/disciples, the end justifies the means.

Science, since Aristotle, formulated hypotheses, gathered preliminary data and information, formulated ways to test the hypothesis, performed experiments, then if there was an indication the hypothesis was correct, the experiments were replicated until the hypothesis became a "Credible Theory".

As time passed, if flaws in the theory were identified, the theory was changed or adjusted to consider the new information. Therein lies the sum of the Fatal Flaws of the "Global Warming Theory". A true scientist modifies the theory, not the data. Hopefully, you comprehend that this renders this system totally unreliable, and it is so on its face.

If you are not convinced at this point, I have a final example that you need to consider. In my opinion this represents the manifestation of the ultimate fatal flaw of the GISTEMP Surface Air Temperature system.

Succinctly stated: The myriad opportunities for bias to flood into the scheme is the ultimate fatal flaw of the scheme. Consider the following.

I am placing two graphs on page 99 and one on page 101. The first graph is the 4-Agency Graph. The second is a published Uncertainty Quantification graph (NASA) published May, 23, 2019. The third is the Berkeley Earth Graph shown before.

Examine these three graphs closely. They all purport to show the global mean anomaly of the earth for nearly the same period of time. Are they close to the same in terms of the global mean anomaly? Let me point out some significant differences.

First: In the Berkeley Earth graph notice that in 1935 the anomaly reached 0.7^{0}C, then decreased to near 0.0^{0}C by 1980. The 4-Agency graph and the Uncertainty graph reached 0.00 for the first time in 1935 and 1938 respectively.

Second: The Berkeley Earth graph exhibits a strong negative correlation between the rising carbon dioxide content of the atmosphere and the anomaly for the period between 1935 and 1980. The other 2 graphs exhibit a strong negative correlation between the rising carbon dioxide content and the anomaly for the period of 1880 to 1910.

Third: All three graphs indicate that the 1980 anomaly was very close to 0.0^{0}C. Between 1880 and 1980 the atmospheric carbon dioxide concentration increased by nearly 43% of the total rise that has occurred between 1880 and 2019. That means that there was essentially no correlation between increasing carbon dioxide and temperature for the first 100 years of the global warming era. How does that happen?

Fourth: Above I wrote: Therein lies the sum of the Fatal Flaws of the "Global Warming Theory". A true scientist modifies the theory, not the data!

The 4-Agency graph shows the declining anomaly between 1880 and 1910. Also note the flat period from 1935 to 1979. From 1998 to 2011 you will note that there is no dip, the subject of the above discussed "climate gate fiasco". Finally, note the rapid rise at the far right after 2011.

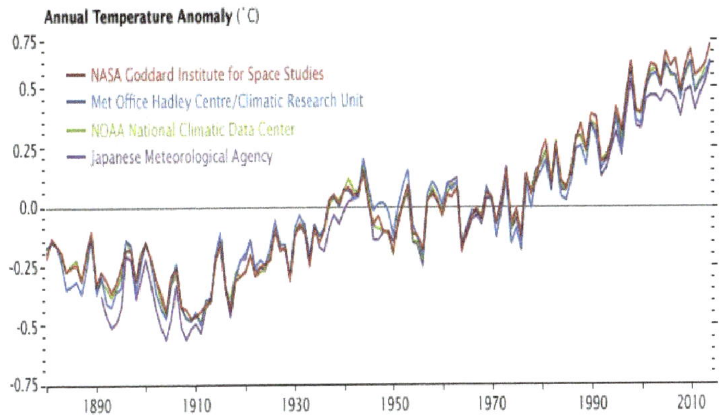

The Uncertainty Quantification Graph:

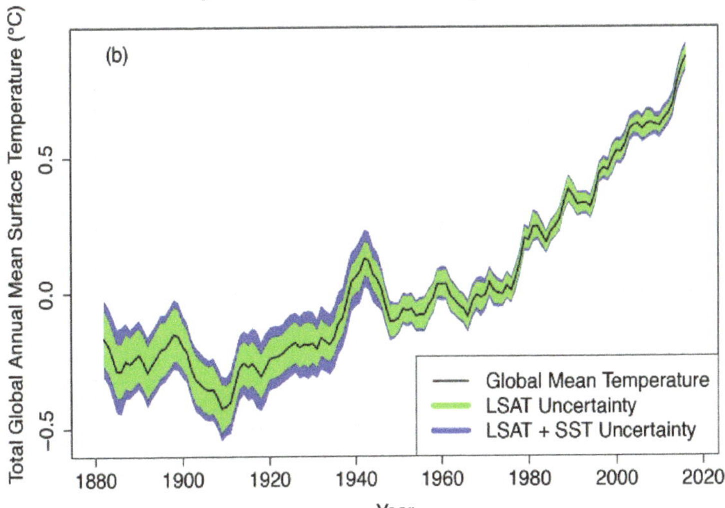

Source: <https://data.giss.nasa.gov/gistemp/uncertainty/>

The Uncertainty Quantification Graph should be essentially the same as the 4-agency graph (to 2014). It is not, because it uses a five year rolling average anomaly.

This technique removes a significant portion of the year to year variation which is due to the four main controlling variables. It also improves the correlation between carbon dioxide concentration and temperature by a factor of 2. It is a method that further changes the data, and completely hides the fact that there was a significant dip during 2017 and 2018.

When the GHCNv4 dataset was adopted, the FAQ sheet contained a statement admitting that some data during the 1951 to 1980 period was "corrected but the effect was minor" (it actually increased the anomaly by 0.2^{0}C, hardly a minor change).

In the same document it also states that the changes increased the 100 year projected increase in the anomaly (a highly significant fact). In this graph (page 99 - Uncertainty Quantification graph) you can see that there is a definite upward slope to the black line in this 30 year "normal" period (equal to about 0.2^{0}C). There was no upward slope in the original data as you can see in the 4-agency graph. This is what is called an internal inconsistency, something that is incompatible with science.

Returning to the discussion of the Berkeley Earth graph on page 101: You see a red line representing the raw data, a blue line representing the corrected data, and a green line representing the fully corrected and homogenized data values. Those lines diverge (differences becomes greater) between 1980 and 2014.

How can that be? During this period the "old fashioned" human, thermometer, pencil, and paper system was replaced with accurate, durable and reliable electronic data acquisition, recording, transmission and storage systems. The errors should have decreased, not increased in numbers and magnitude.

Examine the Berkeley Earth graph below, paying

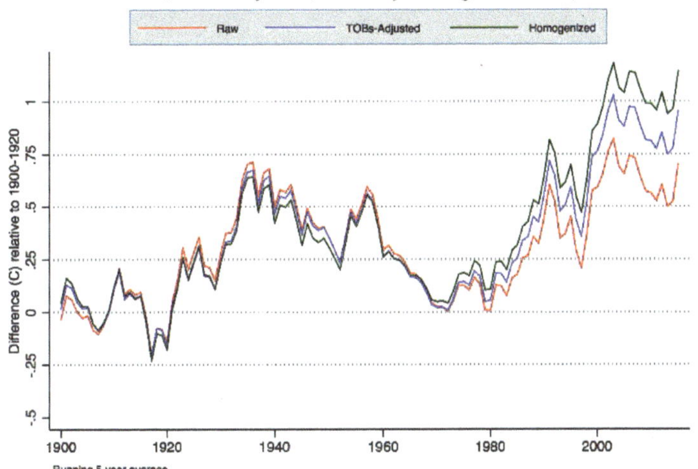

particular attention to the period from 1980 to the end of the graph (2014). Note the differences in the 3 lines.

I have used and relied upon the same type of environmental monitoring systems for many purposes for over 45 years. They were good 45 years ago and have improved greatly over time. This "need" for increasing numbers and/or magnitude of "corrections" flies in the face of reality.

Remember Burl Majors (p. 31) and his description of odiferous occurrences? This does smell like a Limburger Cheese and Salami sandwich that has been stored for two weeks in a warm saddle bag (it reeks)!

Author's note: I have spent over 55 years involved in basic and applied research, some peer reviewing for scientific journals, production agriculture, agricultural management and environmental and agricultural consulting. I have never seen a situation in which it was acceptable to edit data, change data, massage data, or otherwise fudge the data as is being done to support the global warming/climate change theory.

Author's note: The justification for the GISSTEMP Global Surface Temperature system was purported to be a lack of a standardized, accurate, measurement of the mean global surface temperature. You have been able to observe what that has become. It appears, from the above three graphs, that this system does not fill the need.

A Final look:

So, how is the GHCNv4 data set different from the GHCNv3 set?

First, it did reverse the downward trend from the 2018 figure of $0.8^{0}C$ to a 2019 value of $0.9^{0}C$.

Second, the number of temperature reporting sites was increased to 29,000 world-wide (13,000 additional sites were "homogenized" into the new data set).

Third, there were editing changes, and algorithm corrections made to temperatures during the 1951 to 1980 reference (normal) period.

Fourth, the range of temperature editing and correcting procedures was increased.

The addition of 13,000 additional temperature collection sites (81% of the prior total) to the GHCNv4 data set is a suspicious and scientifically indefensible action for the following reasons.

First: This is the second time since 2011 (8 years) that a significant number of temperature monitoring sites were added (homogenized into) to the data set, and both times the result was to reverse a downward trend.

Second: I see a strong suggestion that the above discussed practice referred to as the "Mann Trick" has been adopted by NASA and others. Rhetorical question: Was the vindication of the homogenization technique in the "Climate Gate" fiasco viewed as license by "climate scientists" to adopt an extremely questionable practice?

Third: When the additional temperature monitoring sites were added to the GHCNv4 data set, the data from these sites during the 1951 to 1980 normal period must exactly match (on average) the original sites or they will skew the "normal" global mean anomaly. The FAQ issued when this data set was adopted admits that the additions made do not meet this requirement.

Fourth: The admitted editing of data during the 1951 to 1980 reference period is problematic to the extent, that the entire system can't claim any credibility. (I realize this statement is redundant. I have written it about other "smelly things".

Epilogue/Conclusions

For over 30 years I have been aware of and concerned with, the above fatal flaws of the global warming theory. I have often approached former students, friends, and even relatives, encouraging them to "write the book disclosing these fatal flaws".

I was deeply involved in some volunteer efforts with a global humanitarian organization along with family and industry organizations. I used that as an excuse for myself because I did not have the time.

As I approached others with the above mentioned proposition there was usually a shocked, even fearful response. It usually centered on the thought: I like what I do for a living and you are asking me to give it up, because they knew, that if they told the truth they would be ostracized, shunned, and denied the status of peer in their professions.

When that happened four years ago with a person whom I know I share the same position on this issue, I decided that I would need to "write the book" or it would not be done.

So I embarked on this journey. My first effort lasted about 12 months and resulted in what I now refer to as "Fatal Fails: Scientific Problems with Global Warming" a comprehensive treatment of this topic.

My brother-in-law, who's opinion I value, was given an early copy of the manuscript to review. After several days his response was: Earl, what are you trying to do? Put the sleep-aid business, out of business?

On reviewing that manuscript from a different perspective (with a few naps during the process), I was convinced that very few would care to read such a mass of information. So I set about attempting to reduce the subject to its essentials and making it readable and even interesting. You will be the final judge if I have accomplished that goal.

There are many problems that underlie the Global Warming/Climate Change theory. The first arises because of the basic assumption underlying the theory which is: All the warming of the earth's surface that has taken place since 1880 is the result of increasing carbon dioxide levels in the atmosphere.

To the degree that this assumption is incorrect, the causation of any temperature change is bound to be erroneous because all warming is credited to the impact of the minor greenhouse gases.

In addition to the fatal flaws which result from the underlying assumptions, multiple fatal flaws arise from the methods employed.

First: The "experimental design" of the GISSTEMP anomaly system was insufficient and problematic. Instead of establishing a uniform, precise, reliable and maintainable system of weather monitoring stations, a hodgepodge mixture of weather stations was used. These homogenized systems have significant problems of accuracy, precision, and positioning.

The system characteristics defined in the last sentence of the paragraph immediately above, precludes the estimation of the earth's surface temperature with the precision and accuracy demanded by the anomaly concept.

Second: To compensate for the fatal flaws which result from the poor experimental design and execution, a system of data changes has been adopted which allows bias to flood into the data and the analytical results. It is difficult to imagine a less scientifically sound temperature estimation system than what is currently used.

It requires constant data editing and changes, including periodic drastic system adaptations to maintain the "global warming theory doctrine" in a manner that will appear acceptable to the masses. In the process, science is redefined, and the methods adopted strain credulity, to "prove" preconceived doctrinal conclusions.

Early in this book you were exposed to a simple, easily understood comparison between the single dimensional modeling system used to support what I referred to as a ridiculous assumption, and a multi-dimensional model or concept which would meet the requirements of this system.

I am sure that some thought I was being harsh and judgmental when I wrote that. I suggested that you continue to read and indicated that I would ask at the end of the work if you still felt that way; so . . . . .

Let us view the contents of a Mason Jar through the prism of a single-dimensional interaction model between temperature and rising atmospheric carbon dioxide concentration with the basic assumption that; "Any heating will be the result of carbon dioxide forcing".

We begin with a 1-quart mason jar containing 1 pint (473 g) of water at a temperature of 20°C.

Step A. We add a net of 473 calories of heat as radiant energy from carbon dioxide.

Step B. The result is that the water temperature is now 21^{0}C.

Step C. The Conclusion: The carbon dioxide has caused a 1^{0}C rise in the water temperature.

Step D. We now add a net of 473 calories of heat as sunlight energy.

Step E. The result is that the water temperature is now 22^{0}C.

Step F. The conclusion: The carbon dioxide has caused a 2^{0}C rise in the water temperature.

Does the above make sense? The answer is no. But, remember the controlling assumption: “all heating will be the result of carbon dioxide forcing of the contents of the mason jar” hence the conclusion, no matter how ridiculous: “the heating is all the result of carbon dioxide forcing of the jar’s contents”.

If the heat came from water, magma, or the greenhouse impact of aerosols, the conclusion would be the same, forced by the underlying assumption.

Now let us view the contents of a mason jar through the prism of a multi-dimensional model between temperature and carbon dioxide, with the basic assumption that any heating of the contents of the jar may be the result of energy from the sun, water, magma, aerosols, and/or carbon dioxide.

We begin with a 1-quart mason jar containing 1 pint (473 g) of water at a temperature of 20^{0}C.

Step A. We add a net of 473 calories of heat as radiant energy from carbon dioxide.

Step B. The result is that the water temperature is now 21^{0}C.

Step C. The Conclusion: The carbon dioxide has caused a 1^{0}C rise in the water temperature.

Step D. We now add a net of 473 calories of heat as sunlight energy.

Step E. The result, the water temperature is now 22^{0}C.

Step F. The conclusion: The carbon dioxide has caused a 1^{0}C rise in the water temperature and the sunlight energy has caused a 1^{0}C rise in the temperature of the water.

If the heat added in A, or D above, resulted from setting the jar in warm water, or on cooling magma, or from the greenhouse effect caused by aerosols, the added heat would be credited to the proper source.

That does not happen with a single dimension, myopic modeling system. It is all lumped together under the carbon dioxide forcing moniker.

When it is broken down it is easily understood, and the fatal flaw forced by the ridiculous, underlying assumption is exposed. Now that was not so difficult to understand, was it?

Copies of the book: “Fatal Flaws” will be available in eBook form from Gatekeeper Press and allied distributors of eBooks. Paperback copies of this book,

“Fatal Flaws” are also available from Gatekeeper Press and allied distributors.

If you would like to have a more comprehensive view of this subject, copies of “Fatal Fails: Scientific Problems with Global Warming” will be available in eBook and paperback format by 4/1/2022.

Again I thank you for reading and considering what is written herein. As Professor Ashton often said: “Truth is truth: You are getting there, it’s more complex than you think and it always will be.”

Source of Berkeley Earth Graph used in this work:

Berkeley Earth Graph: Was found in an article entitled “Thorough, not thoroughly fabricated: The truth about global temperature data” in an ars Technica publication @ (https://arstechnica.com/science/2016/01/thorough-not-thoroughly-fabricated-the-truth-about-global-temperature-data/).

Appendix #1: The Normalization of Variance and Groupthink

Why is the global warming theory so widely accepted, and the purveyors of this fraud so widely respected? The global warming theory is a simple answer. It is much too simple!

Following the explosion and breaking apart of the Space Shuttle Challenger, a series of investigations into the cause of this disaster were conducted. In the course of these investigations, a NASA report entitled "The Cost of Silence" was published.

The report pointed out how the normalization of deviance played a role in this disaster. Vaughan's Normalization of Deviance states: "Social normalization of deviance means that people within the organization become so much accustomed to a deviation that they don't consider it as deviant, despite the fact that they far exceed their own rules for the elementary safety". Diane Vaughan, 1996 (from NASA report: "The Cost of Silence": Normalization of Deviance and Group think; Senior Management ViTS Meeting)

The same principles are applicable to the adoption of the global warming doctrine, and the normalization of practices that depart significantly from earlier scientific principles in order to "prove the theory" and then declare it "settled science".

For example, as discussed in this work: It is a fact that the increase in carbon dioxide concentration and a slight increase in the measured mean temperature of the earth have been correlated when the edited data are used. There is also a list of facts that call into question the validity of the theory. They are: 1.The underlaying assumptions, 2.The shoddy experimental design and execution; and 3.The vast extent and array of changes that have been made to the data.

In spite of the above, over time, through widespread promotion and intolerance of dissent, this uncontested declaration that correlation "proved the theory" became widely accepted. In fact, that was not, and is not the case, because correlation is not enough to prove cause and effect (see Appendix #2: A Correlation Primer p. 116).

Yet the declaration of the "proof" of the theory was followed by a period of debate over how strong the "proof" was. Then, because the loudest voices were on the global warming side of the argument, it became "settled science".

Beneficiaries were; academia (with notable exceptions who were silenced), the media (it makes a good story and/or Saturday morning cartoons), politicians who positioned themselves to take advantage of an emotion-based movement, and many who figured it was a tremendous opportunity to accumulate wealth.

The above quote by Vaughn on normalization of deviance comes from a very sad circumstance. The mistakes made and accepted in the lead up to the explosion of the space shuttle shortly after takeoff, resulting in the loss of the entire crew's lives.

The key is the phrase: "people within the organization become so much accustomed to a deviation that they don't consider it as deviant." We all have human failings. Some of the greatest are rationalization, myopic views of what is, and an ego limited ability to recognize when we have been, or are, wrong.

The following: http://www.geocities.ws/oralcomp-groupthink/symptoms.htm; a discussion of "Normalization of Deviance and symptoms of GroupThink" is instructional. Slight adaptation of the 8 symptoms of groupthink in the rocket launch environment, to fit the global warming/climate change situation, should be easy enough for anyone capable of critical thought.

Authors Note: The following is, in my opinion, an explanation of how and why the global warming theory has developed into a massive secular religion that is rapidly making the transition into the position of a state sponsored religion.

It is now evident that the intolerance of dissent is increasingly becoming not just an academic and social force, but that it is being promoted within the educational system as well as in governmental directives, policies, regulation and by social media "platforms".

The current political environment is a manifestation, in many ways, of the 8 symptoms of groupthink. For years we have been observing the normalization of deviance in the name of diversity, which is now being turned on its head into intolerance of anything but one mindset and thought pattern.

Janis lists eight specific symptoms of a group undergoing GroupThink: *The Marker Felt Font is Janis's description of the symptom*; The, As to Symptom #; is my comment.

1. **Illusion of Invulnerability: Experts in a group will feel that nothing can go wrong after making a decision, because they "are the best" and therefore the decision they made cannot go wrong.** As to Symptom #1: Bias is to be avoided at all costs in safety and scientific research. Arrogance leads to the worst kind of bias, because, "We are the best" gives license to bend or ignore the rules.
2. **Belief in Inherent Morality of the Group: "Experts in a group will believe that the decisions they are making will be morally correct".** As to Symptom #2: In the case of global warming/climate change experts, not only do they presume to be morally correct, they are, in their own collective minds, morally superior to the point that they have not only the right, but a duty, to shut down and punish dissent.
3. **Collective Rationalization: Members discount warnings that their thinking may be irrational.** As to Symptom #3: Once the step is taken to attribute all of the warming of the earth to 3 or 4 minor greenhouse gases and you have signed on with all the force of your faith and reputation (or utilized the movement to gain tenure or wealth); you can't admit that your faith and newly strengthened reputation is built on faulty assumptions or incomplete data and facts.
4. **Out-group Stereotypes: The members will all have the same stereotype of the problem.** As to Symptom #4: The current global warming/climate change groupthink has been built over 50+ years of environmental, insti-

tutionalized thought, and yes, even mind control patterns. When you start with theme-cartoons for toddlers, reinforce basic thoughts and patterns of thinking throughout K-12, and then continue with outright indoctrination in college and graduate school, you are guaranteed that a very high percentage of your subjects will have the same thinking patterns. Add the basic components of intolerance of dissent, declare it settled science, and you have accomplished your goal. Very few are capable of anything else. At any point, the question of "what are the weaknesses or problems with this theory" can't be asked or even thought about because it is "settled science," demanding blind faith of everyone (the Prophets[1] can't be questioned).

5. **Self-Censorship: "When a specific expert is asked what to do about a problem, the expert will not give an answer directly contradicting the group's wishes".** As to Symptom #5: After all, it is "settled science" and you have witnessed what happens to dissenters, they are defamed, persecuted and destroyed.

6. **Illusion of Unanimity: Even if some individuals of the group privately do not consent to the decision made, the individual will not speak out. The individual's silence will then be perceived as consent.** As to Symptom #6: No explanation required. We have witnessed the cost of dissent.

7. **Direct Pressure on Dissenters: If an individual does speak out against the will of the majority of the group, all other members will attempt to**

quell the "dissident". As to Symptom #7: I remember Dr. Gray of the National Hurricane Center and a number of others. You may not remember them, they were ostracized, defamed, denied grants and promotion, and/or terminated; in short, they were destroyed because they knew and defended the truth.

8. **Self-Appointed Mindguards/**Prophets[1]**: The leader of the group of experts will refuse to hear any argument contradicting his own wishes or the will of the majority.** As to Symptom #8: Self-Appointed Mindguards: We have witnessed this process grow and expand to where the numbers of "leaders" and their reactions to any argument contradicting their own is vast in extent and extreme in the intensity of intolerance.

Add to the eight symptoms of groupthink the economic and power benefits that flow to the "leadership class" and you have a self-perpetuating, action, mind, and thought control system, that the Stalinists and Maoists of the 1940s, 50's, 60's and 70's could only have dreamed about. Add to this Google's social behavioral monitoring abilities (referring to what they have developed for China) and you have George Orwell's 1984 on steroids; by 2025?

[1] You will note that I have used certain specific religious designations applied to the leaders and followers of the global warming/climate change religion. I mean no disrespect toward any religious authority who sincerely believes and teaches the doctrine of a true religion (i.e., one who believes in and worships a

higher power than man). I do feel that self appointed leaders of a secular religion, designed for nefarious purposes, do not deserve such respect.

APPENDIX #2

A "Correlation Primer"

A mathematical analysis technique called "Correlation Analysis" is used to determine the degree of correlation which exists between two variables. The degree of correlation is expressed as a value of "r". The value of r can then be used to estimate the coefficient of determination "r^2" which is an estimate of how much of the observed effect is attributable to the variables being examined.

The "r" value may be positive if one factor goes up when the other goes up, or it may be a negative if, one factor goes up and the other one goes down. For example, my weight is positively correlated to how many calories I consume each day. If I have a low level of physical activity, the r value is positive and very close to 1.0 (the + sign is usually not displayed, it is understood).

Remember the Allegory about My Strict Diet in Chapter 2 (p. 16). If I considered my initial inferred assumption, that any weight loss would be attributable to the calories I consumed to be settled science, I would end up attributing my weight loss to my higher calorie intake. This is an example of how a single dimensional model (comparing my weight loss to my increased calorie intake) would mislead me, if my basic assumption was that any weight loss would be the result of my increased calorie intake.

If I examined the results of the above example closely, I would find that: First; there was a strong negative correlation between my increased calorie consumption and weight loss but my conclusion would be fatally flawed. Why? Because: Second; there is no definable, known mechanism that would explain why a greatly increased calorie intake could cause weight loss; and Third; the results would not be repeatable (without the accompanying exercise).

The above paragraph defines what is known as Jacob's Rules. In order to demonstrate a cause and effect relationship between two variables there are three rules that must be satisfied: First; there must be a correlation (may be positive or negative). Second; there must be a definable, proven mechanism by which a change in one variable may cause the other variable to change, and Third; the results must be repeatable.

The foundational assumption in the allegory of My Strict Diet is fatally flawed because it fails to meet the second and third requirements set forth by Jacob's Rules.

In this work I have made the case that the inferred correlation between a rising CO_2 concentration in the atmosphere and an increasing earth's surface temperature only appears to be strong if you do not run the proper mathematical analysis.

If you did the proper mathematical analysis, using non-fudged data, (the raw data displayed by the red line in the Berkeley Earth Graph shown several times) and examined closely the other requirements, the invalid nature of the basic assumptions are obvious. When the impacts of the 4 major drivers of earth's surface temperature are properly credited as causative forces, the picture is vastly different and clear.

Let me say that again. When the impacts of the 4 major drivers of earth's surface temperature are properly credited as the main causative forces, the picture is vastly different. Failure to treat nature as the beautiful, dynamic, system it is, results in fatal fails.

Appendix #3:

The Ultimate Fatal Flaw of the Global Warming Theory!

As I was making final preparations to publish this work and was struggling to find simpler ways to say or illustrate certain principles, I came upon the following two definitions.

1. Climate forcing: An energy imbalance imposed on the climate system either externally or by human activities.

2. Direct radiative forcing: A climate forcing that directly affects the radiative budget of the Earth's climate system; for example, added carbon dioxide (CO2) absorbs and emits infrared radiation.

When I viewed these 2 definitions next to each other I was reminded of a conundrum (a confusing and difficult problem or question; the cooling effect of increased carbon dioxide in the atmosphere mentioned on p. 56 of this work).

The conundrum, simply stated is: Climate scientists understand that the sunlight that reaches the earth's surface warms the earth. The warm earth releases some of that heat (25% of the energy that leaves earth's surface) as infrared "light" or radiation away from the surfaces.

Some of the radiant energy released from the earth's surface is intercepted by carbon dioxide, methane and other minor greenhouse gases. This heat is then re-radiated or released, some toward the earth and some toward space.

This interception of radiant energy from the earth's surface and the re-radiation of some of it back to the earth is the contribution of the minor greenhouse gases to the earth's greenhouse effect.

NASA and the UN's IPCC (Intergovernmental Panel on Climate Change) has determined that the net amount of excess heat retained by the earth's surface during the last 140 years because of the increased greenhouse impact is between 0.70 to 1.0 Wm^2 (watts per square meter).

The claim is: The retention of that heat is what has caused the approximate 1^oC increase in the earth's temperature. <u>If one is used to uni-dimensional models and thinking patterns that is "the end of the story".</u>

If one understands that it is not that simple, they realize that there are several additional chapters to the story. These other chapters are what is discussed in this book "Fatal Flaws". **Attention:** Here is the point of this Appendix!

Not being bound to a uni-dimensional thought process, I returned to an examination of what I have written and realized that there is still another factor or force which has escaped examination (by everyone).

Climate scientists stress how efficient the minor greenhouse gases are at absorbing radiant energy. The last phrase in the 2nd definition above is: "for example, added carbon dioxide (CO2) absorbs and emits infrared radiation." The question then is: If you increase the concentration of these gasses in the atmosphere what happens when they absorb sunlight moving toward the earth? This question has not been considered by global warming/climate change advocates. I was aware that this conundrum exists but until I observed the two definitions above side by side I had not bothered to consider it in-depth. When I did I realized that:

We know the answer to that question. The minor greenhouse gasses absorb some of the suns radiant energy as it moves toward the earth's surface. The "excited carbon dioxide molecules" then re-radiate it; some toward the earth and some back into space (the reverse of the greenhouse effect; it results in cooling of the earth, because less of the sunlight energy reaches the earth's surface).

That leads to a second question which is: Given that the intensity of the radiant infrared energy moving toward the earth is much higher than the intensity of the radiation given off by earth's surfaces, what is the net effect of the increased concentrations of the minor greenhouse gasses?

I will predict that you will never see the question above discussed or answered anywhere but here because:

Assuming the minor greenhouse gases are as effective as claimed in intercepting radiant energy, the net impact of the increased concentrations would be a significant cooling of earth's surface. This is because the incoming infrared irradiance intensity is much higher than the outgoing energy level (there is more energy to intercept and radiate back into space).

It must always be remembered that the "anomaly system or concept" results in a maximum of only 1.1^{0}C increase in the earth's surface temperature (using the maximum fudged figures). When you are dealing with 1/10, or a few tenths of a degree, the margins for error and bias are very slim.

This may be a little technical but it needs to be said (I will simplify it as much as possible). The intensity of the incoming solar irradiance in the infrared portion of the spectrum is greater, by at least ten times, or more, than the intensity of the outgoing radiant energy. That means that the cooling impact of these gasses is greater than the warming impact! (See: "Fatal Fails" for a complete discussion of this Fatal Flaw of the Global Warming/Climate Change Theory)

P.S.

According to the EPA's latest figures the US total releases of greenhouse gasses was equal to 7.2 billion metric tons in 2007 and decreased steadily to 6.6 billion metric tons in 2020.

The Global emissions of these gasses was 40 billion metric tons in 2005 and increased to 47 billion metric tons in 2020.

The above figures tell us that the US produces 14.4% of the globally released greenhouse gasses (much lower than what is claimed by Global Warming advocates). China passed the 29% level of global greenhouse gas emissions recently.

A final Rhetorical Question: What is the potential impact of the US spending tens of trillions of dollars in "tearing it all down and building it back better" if China, India, Russia and most of the rest of the world continues to increase their emissions of greenhouse gasses?

In the book: "Fatal Fails: Scientific Problems With the Global Warming Theory" I discuss the negative environmental impacts of the proposed "Build it Back Better," pending disaster.

www.ingramcontent.com/pod-product-compliance
Ingram Content Group UK Ltd.
Pitfield, Milton Keynes, MK11 3LW, UK
UKHW062257290726
14090UKWH00017B/740